Functional Equations with Causal Operators

Stability and Control: Theory, Methods and Applications
A series of books and monographs on the theory of stability and control
Edited by A.A. Martynyuk
Institute of Mechanics, Kiev, Ukraine
V. Lakshmikantham
Florida Institute of Technology, USA

Please see the back of this book for other titles in the Stability and Control: Theory, Methods and Applications series.

Functional Equations with Causal Operators

C. Corduneanu

CRC Press
Taylor & Francis Group
Boca Raton London New York

CRC Press is an imprint of the
Taylor & Francis Group, an **informa** business
A TAYLOR & FRANCIS BOOK

First published 2002 by Taylor & Francis

Published 2021 by CRC Press
Taylor & Francis Group
6000 Broken Sound Parkway NW, Suite 300
Boca Raton, FL 3487-2742

© 2002 by Taylor & Francis Group, LLC
CRC Press is an imprint of Taylor & Francis Group, an Informa business

No claim to original U.S. Government works

ISBN-13: 978-0-415-27186-8 (hbk)

Typeset in 10/12 pt Times New Roman by
Newgen Imaging Systems (P) Ltd, Chennai, India

Visit the Taylor & Francis Web site at
http://www.taylorandfrancis.com

and the CRC Press Web site at
http://www.crcpress.com

British Library Cataloguing in Publication Data
A catalogue record for this book is available from the British Library

Library of Congress Cataloging in Publication Data
A catalog record for this book has been requested

Contents

Introduction to the Series

The problems of modern society are both complex and interdisciplinary. Despite the apparent diversity of problems, tools developed in one context are often adaptable to an entirely different situation. For example, consider the Lyapunov's well known second method. This interesting and fruitful technique has gained increasing significance and has given a decisive impetus for modern development of the stability theory of differential equations. A manifest advantage of this method is that it does not demand the knowledge of solutions and therefore has great power in application. It is now well recognized that the concept of Lyapunov-like functions and the theory of differential and integral inequalities can be utilized to investigate qualitative and quantitative properties of nonlinear dynamic systems. Lyapunov-like functions serve as vehicles to transform the given complicated dynamic systems into a relatively simpler system and therefore it is sufficient to study the properties of this simpler dynamic system. It is also being realized that the same versatile tools can be adapted to discuss entirely different nonlinear systems, and that other tools, such as the variation of parameters and the method of upper and lower solutions provide equally effective methods to deal with problems of a similar nature. Moreover, interesting new ideas have been introduced which would seem to hold great potential.

Control theory, on the other hand, is that branch of application-oriented mathematics that deals with the basic principles underlying the analysis and design of control systems. To control an object implies the influence of its behavior so as to accomplish a desired goal. In order to implement this influence, practitioners build devices that incorporate various mathematical techniques. The study of these devices and their interaction with the object being controlled is the subject of control theory. There have been, roughly speaking, two main lines of work in control theory which are complementary. One is based on the idea that a good model of the object to be controlled is available and that we wish to optimize its behavior, and the other is based on the constraints imposed by uncertainty about the model in which the object operates. The control tool in the latter is the use of feedback in order to correct for deviations from the desired behavior. Mathematically, stability theory, dynamic systems and functional analysis have had a strong influence on this approach.

Volume 1, *Theory of Integro-Differential Equations*, is a joint contribution by V. Lakshmikantham (USA) and M. Rama Mohana Rao (India).

Volume 2, *Stability Analysis: Nonlinear Mechanics Equations*, is by A.A. Martynyuk (Ukraine).

Volume 3, *Stability of Motion of Nonautonomous Systems: The Method of Limiting Equations*, is a collaborative work by J. Kato (Japan), A.A. Martynyuk (Ukraine) and A.A. Shestakov (Russia).

Volume 4, *Control Theory and its Applications*, is by E.O. Roxin (USA).

Volume 5, *Advances in Nonlinear Dynamics*, is edited by S. Sivasundaram (USA) and A.A. Martynyuk (Ukraine) and is a multiauthor volume dedicated to Professor S. Leela.

Volume 6, *Solving Differential Problems by Multistep Initial and Boundary Value Methods*, is a joint contribution by L. Brugnano (Italy) and D. Trigiante (Italy).

Volume 7, *Dynamics of Machines with Variable Mass*, is by L. Cveticanin (Yugoslavia).

Volume 8, *Optimization of Linear Control Systems: Analytical Methods and Computational Algorithms*, is a joint work by F.A. Aliev (Azerbaijan) and V.B. Larin (Ukraine).

Volume 9, *Dynamics and Control*, it edited by G. Leitmann (USA), F.E. Udwadia (USA) and A.V. Kryazhimskii (Russian) and is a multiauthor volume.

Volume 10, *Volterra Equations and Applications*, is edited by C. Corduneanu (USA) and I.W. Sandberg (USA) and is a multiauthor volume.

Volume 11, *Nonlinear Problems in Aviation and Aerospace*, is edited by S. Sivasundaram (USA) and is a multiauthor volume.

Volume 12, *Stabilization of Programmed Motion*, is by E.Ya. Smirnov (Russia).

Volume 13, *Advances in Stability Theory at the end of the 20th Century*, is edited by A.A. Martynyuk.

Volume 14, *Dichotomies and Stability in Nonautonomous Linear Systems*, is by Yu.A. Mitropolskii, A.M. Samoilenko and V.L. Kulik.

Volume 15, *Almost Periodic Solutions of Differential Equations in Banach Spaces*, is by Yoshiyuki Hino, Toshiki Naito, Nguyen Van Minh and Jong Son Shin

Volume 16, *Functional Equations with Causal Operators*, is by C. Corduneanu.

Due to the increased interdependency and cooperation among the mathematical sciences across the traditional boundaries, and the accomplishments thus far achieved in the areas of stability and control, there is every reason to believe that many breakthroughs await us, offering existing prospects for these versatile techniques to advance further. It is in this spirit that we see the importance of the "Stability and Control" series, and we are immensely thankful to Taylor and Francis for their interest and cooperation in publishing this series.

Preface

This book is dedicated to the investigation of functional or functional differential equations involving causal operators. These operators are also called nonanticipative, or abstract Volterra operators. The term "causal" is prevalent in the engineering literature.

The definition of causal operators is very simple: an operator V, acting on a given function space $E([0, T], \mathbb{R}^n)$, is called *causal*, if for any pair of function x, y of E, such that x and y coincide on an interval $[0, t]$, $t \le T$, Vx and Vy also coincide on that interval. In other words, the values of Vx up to a given point t are determined only by the values taken by the function x on the interval $[0, t]$.

The idea of considering such operators appears implicitly in Volterra's work, but a sharp definition and further consideration appear in the paper of L. Tonelli [1]. In this paper, the functional equation

$$x(t) = f(t) + A(t, x_0^t(s))$$

is considered, where the second term in parantheses means the restriction of the function x to the interval $[0, t]$. The notation is obviously inspired by Volterra's work, and the above equation reminds us instantly of the Volterra integral equation

$$x(t) = f(t) + \int_0^t k(t, s, x(s)) \, ds.$$

Tonelli's paper was dedicated to proving the existence and uniqueness of the solution of the functional equation he has devised by means of causal operators. The equation has been investigated in the space of continuous functions and the hypotheses are formulated in such a way that the compactness of the operator A is assured. This result of Tonelli is, very likely, the first existence result for equations with general causal operators.

The next significant step in developing the theory of functional equations involving cauasal operators was made in 1938 by A.N. Tychonoff [1]. The definition given by Tychonoff to causal operators is as formulated above, and besides the existence of solutions the importance of these types of operators or equations for other fields is emphasized.

In retrospect, it may appear somewhat strange that the concept of a causal (or, as both Tonelli and Tychonoff call it, Volterra) operator did not attract the immediate attention of researchers. A possible explanation may be that at the time this concept was advanced, the relatively new methods of functional analysis did not constitute the main tool of many investigators.

Gradually, the theory of functional equations with causal operators has caught the attention of researchers. From the 1960s we mention the papers by R. Driver [1], C. Corduneanu [2], and Z.B. Caljuk [1]. The last-quoted paper deals with functional inequalities with causal operators.

In the 1970s there have been many authors dealing with causal operators/equations. For the first time in book form, these operators are discussed in the book by L. Neustadt [1]. Journal papers have been published by M. Kwapisz and his coworkers, by V.G. Kurbatov, L.A. Zhivotovskii, S. Szufla (see the references under these names). A group of researchers from Russia (Perm Technical University), under the leadership of N.V. Azbelev, has started the systematic investigation of linear functional differential equations with causal operators. The activity of this group continues nowadays, with its members in Russia, Israel and other countries. Also in the 1970s, a good deal of research work has been conducted with regard to the equations with infinite delay. The books by J.K. Hale [1] and A.D. Myshkis [1] have also contributed to making the subject of causal operators an attractive topic for researchers.

In the 1980s, the theory of functional or functional differential equations with causal operators made serious steps towards its maturity. A large number of papers were published during this decade in the USA, in the former Soviet Union, Italy and other countries. Most of the basic problems of this theory, including stability theory, approximation procedures and other aspects, have been considered by many authors. Our list of references provides a large number of items from that period, without any claim to being complete. It is also important to notice that during the 1980s, a large number of engineering papers (system science) were produced. The books by W.J. Rugh [1] and M. Schetzen [1] cover some of these topics. Fundamental results concerning causal general operators have been obtained by I.W. Sandberg [1], [2], [3], not necessarily related to the theory of functional equations.

The 1990s were characterized by an increasing interest in the theory of functional equations with causal operators, often related to applications. Several books including results concerning functional equations with causal operators have been published, or are in an advanced state of publication: G. Gripenberg, S.O. Londen and O. Staffans [1], C. Corduneanu [10], N.V. Azbelev, V.P. Maksimov and L.F. Rakhmatullina [1], N.V. Azbelev and P.M. Simonov contain chapters or conspicuous sections dedicated to this theory.

The journal literature is currently growing steadily, with at least one periodic publication called *Functional Differential Equations* being mostly dedicated to this theory. A serial publication with the same title has been published by N.V. Azbelev and his collaborators in Perm. Many other journals include contributions on this subject, authored by researchers from Russia, Ukraine, Israel, the USA, Italy, Ireland, Japan, Georgia, Australia, Poland, Germany and other countries.

This book evolved during the period 1991–1998, when the author and his former students held a weekly seminar at the University of Texas at Arlington. In particular, Dr Mehran Mahdavi and Dr Yizeng Li were active and wrote their PhD theses about functional differential equations with causal operators. After they graduated and went on to teach at other institutions of higher learning, the cooperation between us continued without interruption. The material contained in this book is in greatest part based on the work we developed separately or jointly. Our list of references is complete with respect to the contributions we have made to this theory, including some applications.

Some of the topics treated in this book have also been the object of some graduate courses the author has taught at the University of Texas of Arlington, starting in the late 1980s until his retirement in 1996. These courses had various titles, and we will mention here those on applied nonlinear analysis, an introduction to control theory, advanced applied differential equations.

One of the aims while teaching such topics was to present a unified treatment of existence theory, as well as other aspects of the theory of functional differential equations, including ordinary differential equations, equations with delayed argument (both finite and infinite), integral equations of Volterra type and integrodifferential equations involving Volterra integral operators. Indeed, the theory of functional differential equations with causal operators has very powerful unifying qualities – a feature we have attempted to illustrate in this book. As a matter of fact, our involvement in the theory of functional equations with causal operators was motivated by the needs of the teaching process, especially for engineering and science graduate students.

Briefly, the structure of the book is as follows: Chapter 1 is an introduction to the concept of functional equations, in general, with particular concern for equations with causal operators; Chapter 2 contains some auxiliary material, mostly pertaining to functional analysis (both linear and nonlinear), as a preparation of the reader for understanding the rest of the material presented in the book; Chapter 3 deals entirely with the existence theory for functional or functional differential equations with causal operators, emphasizing the fact that these equations contain as particular cases several classes of functional equations encountered in the literature; Chapter 4 is dedicated to the theory of linear and quasilinear equations with causal operators, including the global character of existence results and the integral representation of solutions; Chapter 5 contains an introduction to stability theory for equations with causal operators, featuring both the method of "first approximation" and the "comparison method" based on Liapunov functionals and differential inequalities; Chapter 6 is completely dedicated to the theory of neutral functional equations with causal operators, and it is based on the work of the author and of M. Mahdavi; Chapter 7 contains some applications of the general theory to problems in optimal control, some generalizations of the existence results by means of Leray–Schauder principle, as well as a review of certain results available in the literature.

In our view, the book is addressed to graduate students in science and engineering, whose scientific horizon should extend beyond the classical theory of ordinary differential equations.

In order to be able to adequately describe phenomena whose evolution is sensibly influenced by their past, we need a theory of functional equations with causal operators, with all its basic constituents: existence, uniqueness, approximation, continuous dependence with respect to data, stability, global behavior and relationships with other problems such as control theory, mechanics of materials with memory. It is our hope that this goal is partially fulfilled by this book. There certainly remains much more to be done in the future.

For the preparation of this book I am indebted to my young colleagues Dr Mehran Mahdavi, Dr Yizeng Li and Dr Zephirinus Okonkwo, who actively participated in the seminar organized at the University of Texas at Arlington during the period 1990–1994. Their interest in this field of research has stimulated all of us in carrying out the work on the problems presented in this book. Occasionally, we had other participants to the seminar, such as Dr Yoshihiro Hamaya from the Okayama University of Science, who spent one semester with us and who presented some of his results on integrodifferential equations.

Finally, I take the opportunity to express my thanks to Mrs Elena Cosma from the Department of Mathematics at the University of Iasi, and to Mrs Marjorie Nutts from the Department of Mathematics at the University of Texas at Arlington, for their support in preparing the manuscript.

My thanks are also directed to Ms Janie Wardle from Gordon & Breach, who has shown constant interest during the completion of this project.

1 Introduction

1.1 Classes of functional equations

The physical sciences and engineering offer a wide array of functional equations, such as ordinary differential equations, partial differential equations, integral equations, integrodifferential equations, equations with delayed argument, and equations which involve combinations of the above-mentioned types. The use of functional equations can be traced back to the beginnings of various physical theories, their age being almost the same as the age of calculus (Newton dealt in depth with differential equations).

Much later in the historical development of the theory of functional equations, other fields of knowledge such as biology (population dynamics), economics (renewal of resources) or social sciences (investigation of processes taking place in society) have started using functional equations as auxiliary tools.

By a *functional equation*, in a broad sense, we understand an equation in which the unknown is a function. All types mentioned above are illustrations of this concept. Several examples, together with their interpretations, will be considered in this chapter, as well as the connections between them.

We shall deal with classes of functional equations for which the unknown function depends on a single variable, usually denoted by t (time). The unknown function will take its values in a finite-dimensional space, usually the Euclidean space $\mathbb{R}^n$. Hence, $x = x(t) = \mathrm{col}(x_1(t), x_2(t), \ldots, x_n(t))$, with all $x_i(t), i = 1, 2, \ldots, n$, scalar (real valued).

Before we start discussing in some detail various classes of functional equations, as well as their interrelations, we shall list here the most common types we shall encounter in our exposition.

The *ordinary differential equation*

$$\dot{x}(t) = f(t, x(t)) \tag{1}$$

is the most often investigated in the literature, due to its simplicity with respect to other types of functional equations and to its numerous applications in mechanics and other fields. When $n = 1$, we have a scalar equation while for $n > 1$ we have a vector equation. Sometimes equation (1) is written as a system of scalar equations, namely

$$\dot{x}_i(t) = f_i(t, x_1(t), x_2(t), \ldots, x_n(t)), \quad i = 1, 2, \ldots, n. \tag{2}$$

Generally speaking, an equation of the form (1) admits infinitely many solutions. The most elementary example in this regard is the equation $\dot{x}(t) = f(t) =$ known, which admits the solutions

$$x(t) = C + \int^t f(s)\, ds,$$

with C an arbitrary constant (real number or vector). A unique solution to (1) is determined by what is usually known as an initial value condition

$$x(t_0) = x^0 \in \mathbb{R}^n, \tag{3}$$

where t_0 is a fixed value of the independent variable t (initial time).

The theory of equations of the form (1), under initial condition (3), is treated in numerous books: E.A. Coddington, N. Levinson [1], Ph. Hartman [1], G. Sansone and R. Conti [1], to mention just a few of them.

The equation (1) expresses the fact that the rate of change of $x(t)$ is dependent (as described by the function f) on the moment t, and the unknown function itself at that moment. Since investigating the evolution of systems and phenomena requires knowledge of the law/rule of change, it means that equation (1) could be regarded as expressing that law/rule.

Sometimes the situation is more complex. For instance, the rate of change at the moment t may depend on something other than t, and $x(t)$. A possibility is to depend also on $x(t-h)$, with $h > 0$ a fixed number. In this case, the rate of change of $x(t)$ will be determined by an equation of the form

$$\dot{x}(t) = f(t, x(t), x(t-h)). \tag{4}$$

Equation (4) says that the rate of change at the moment t also depends on the status of the system at a past moment $(t - h)$. For instance, if $x(t)$ designates the number of individuals in a certain population, then (4) expresses the fact that the rate of change at the moment t is dependent on the number of individuals in that population, at the moment $t - h$. Actually, only the individuals able of reproduction should be counted, but their number is usually a fraction of $x(t - h)$.

Functional (differential) equations of the form (4) are thoroughly investigated in the mathematical and engineering literature. They are known under various names: *equations with delayed argument* or simply *delay–differential equations*, or *equations with retarded argument*.

The theory of equations with delayed argument is treated in several books and monographs, among which we mention A.D. Myshkis [1], A. Halanay [1], J.K. Hale [1].

Functional (differential) equations more general than (4), but also belonging to the wide category of delay equations, have been also investigated by many authors, particularly by J.K. Hale [1] and his followers. They are written in the form

$$\dot{x}(t) = f(t, x_t), \tag{5}$$

where $x_t(s) = x(t + \tau), -h \leq s \leq 0$. In other words, $f(t, u)$ depends on t, but also on the "functional argument" u. One assumes that u belongs to a certain function space $E([-h, 0], \mathbb{R}^n)$, which consists of maps of the interval $[-h, 0]$ into $\mathbb{R}^n$. The cases $E = C$ or $E = L^2$ are often considered.

The main distinction between (4) and (5) resides in the fact that the right-hand side in (4) is a function in the usual (classical) sense, depending upon a finite number of real variables, while the right-hand side of (5) is a *functional*, i.e., a function (a map) in which the independent variable is itself a function. For instance, the equation

$$\dot{x}(t) = \int_{-h}^{0} k(x) x_t(s) \, ds \tag{6}$$

is such an example, the variable x_t being a function belonging to a given function space.

For the delay equations like (4), (5) or (6), the initial value problem is different from the case of equation (1), namely, one must assign an initial datum which is a function defined on $[-h, 0]$:

$$x(s) = x_0(s), \quad s \in [-h, 0], \tag{7}$$

with $x_0 \in E([-h, 0], \mathbb{R}^n)$.

For the theory of functional differential equations of the form (6), the following reference is pertinent: J.K. Hale [1]. Of course, there is a vast journal literature on this subject.

Another type of functional equation that has been dealt with in the literature, starting with the nineteenth century, has the form

$$x(t) = f(t) + \int_{0}^{t} K(t, s, x(s)) \, ds, \tag{8}$$

and it is known as a Volterra-type *integral equation*. The functions f and k are given functions (scalar or vector valued), while $x(t)$ is the unknown. Such equations do appear in many applied problems, and their study has been under researchers' attention for more than a century. The following references cover most of the aspects encountered in the classic literature: V. Volterra [1], T. Lalescu [1], R.K. Miller [1], A.C. Pipkin [1]. Unlike the theory of ordinary differential equations of the form (1), the integral equations of the form (8) have known a flourishing development only relatively recently (the last 30–35 years).

Another class of integral equations, known as Fredholm-type equations, is represented by

$$x(t) = f(t) + \int_{0}^{T} K(t, s, x(s)) \, ds, \tag{9}$$

with $T > 0$ a fixed real number. The linear case of (9) has conducted Fredholm to the celebrated theory, which is at the origin of modern *spectral theory* (in functional analysis). For some results concerning equations of the form (9) and related forms, see P.P. Zabreyko *et al.* [1], C. Corduneanu [10].

Based on the types of equations briefly described above, one can construct various mixed types of functional equations which are present in the contemporary mathematical or applied science literature. For instance, *integrodifferential equations* of the form

$$\dot{x}(t) = f\left(t, x(t), \int_{0}^{t} K(t, s, x(s)) \, ds\right) \tag{10}$$

have been investigated by many authors. It is interesting to point out the fact that an initial value condition of the form (3) is usually sufficient to determine a unique solution to (10), say $x = x(t)$, on an interval $[0, a], a > 0$. See, for instance, V. Lakshmikantham and M.R. Mohana Rao [1], where both local and global existence results are provided, as well as further references.

As far as the interpretation of an equation like (10) is concerned, let us consider the special case

$$\dot{x}(t) = f(t, x(t)) + \int_0^t k(t, s)x(s)\, ds, \tag{11}$$

which obviously appears as a "perturbed" equation for (1). The perturbing term $\int_0^t k(t, s)x(s)\, ds$ can be regarded as a "memory" term. In other words, the past values of $x(t)$, $x(s)$ for $0 \leq s < t$, influence the evolution of the process. Of course, (11) can be also regarded as a delay equation.

Many other types of integrodifferential equations have been investigated during the past century. We indicate here some of the best known sources: V. Volterra [1], V. Lakshmikantham and M.R. Mohana Rao [1].

1.2 Equations with causal operators

We shall briefly describe in this section the functional equations that will constitute the main object of investigation of this book. Of course, the concept of *causal operator* has to be defined first.

Let $E = E([0, T), \mathbb{R}^n)$ be a function space. Its elements are maps from the interval $[0, T)$ into $\mathbb{R}^n$, and a convergence kind is usually associated to E (for sequences of elements/functions in E). Examples will be considered in subsequent chapters.

Let $V : E \to E$ be a map, commonly called an *operator* on E. We shall say that V is a *causal operator*, or *nonanticipative*, if the following property holds: for each couple of elements of E, such that $x(s) = y(s)$ for $0 \leq s \leq t$, there results $(Vx)(s) = (Vy)(s)$ for $0 \leq s \leq t$, with $t < T$ arbitrary.

The above definition of causal operators is, very likely, due to L. Tonelli [1]. It is true that V. Volterra came very close to this concept, a reason that inspired Tonelli to designate causal operators as *Volterra operators*. Since Volterra operators, in the classical sense, are integral operators (see the right-hand side of equation (8)), we thought the best term to be used is *abstract Volterra operators* (see C. Corduneanu [10], for instance).

The definition of causal operators should be slightly modified when E is a space of measurable function on $[0, T)$. Namely, instead of $x(s) = y(s)$ on $[0, t]$, we should understand the above inequality in the sense of measure theory, i.e., valid almost everywhere on $[0, t]$.

Some examples of causal operators are provided by

$$(Vx)(t) = \int_0^t K(t, s, x(s))\, ds, \quad t \in [0, T), \tag{12}$$

where $K(t, s, x)$ is a function with values in $\mathbb{R}^n$, defined for $0 \leq s \leq t < T$ and $x \in \mathbb{R}^n$. For instance, under the continuity assumption on K, V is defined for every continuous $x(t)$. A more general situation can be envisaged, substituting measurability for continuity.

Instead of the usual integrals (Riemann–Cauchy or Lebesgue), one could use other measures. For instance, considering the linear case, the operator

$$(Lx)(t) = \int_0^t x(s)\, d_s\mu(t, s),$$

under adequate conditions on the family of measures $\mu(t, \cdot)$, is also an operator which satisfies the causality condition.

We shall discuss in more detail the concept of causal operator in Chapter 2. Now, we want to illustrate the generality of this concept, and how the classical types of operator equations encountered in Section 1.1 are covered by this framework.

First, if we compare the ordinary differential equation (1) with the causal differential equation

$$\dot{x}(t) = (Vx)(t), \quad t \in [0, T), \tag{13}$$

under initial condition (3), we see that, for (1), the operator V of (13) has to be chosen as

$$(Vx)(t) = f(t, x(t)). \tag{14}$$

Such operators are obviously causal. They can be represented by a function of real variables $(t, x) \in \mathbb{R} \times \mathbb{R}^n$. In the literature they are known under the name "Niemytskii operators". Sometimes they are called "Carathéodory operators", even though it's not clear that Carathéodory had envisaged the problem from an operatorial point of view.

If we consider now the equation with delayed argument (4), we see that V in (13) must be chosen as

$$(Vx)(t) = f(t, x(t), x(t - h)), \quad t \in [0, T). \tag{15}$$

From (14) it clearly appears that in order to be able to define the operator on the space $E([0, T), \mathbb{R}^n)$, one must know the values of $x(t)$ on $[-h, 0]$, in order to give sense to the formula (15). This is exactly what condition (7) provides. As pointed out above, for an equation with delayed argument the initial condition involves an "initial function".

Let us move now to the integral equation (8), which is known as a Volterra integral equation (of the second kind). Since no differentiation is involved in (8), a direct comparison of (8) and (13) is not immediately possible.

We will consider now the functional equation

$$x(t) = (Vx)(t), \quad t \in (0, T), \tag{16}$$

which has more features in common with the Volterra equation (8).

Let us point out that, in his pioneering paper on functional equations with causal operators, L. Tonelli [1] dealt exactly with the equation (16), which he called a *functional equation of Volterra*.

The comparison of (8) and (13) is now possible, and we need to take in (13):

$$(Vx)(t) = f(t) + \int_0^t K(t, s, x(s)) \, ds. \tag{17}$$

From (17) we see that V is causal. Hence, developing a theory of functional equations with causal (abstract Volterra) operators, we will cover also the theory of equations of the form (8), also with causal operator.

It is adequate, at this point, to compare the functional differential equation (13) with the functional equation (16) in which V stands for a causal operator (of course, not the same in both equations). If conditions allow – and we shall always work under such conditions – one can integrate both sides of (13) from 0 to t, $0 < t < T$, and obtain the equation

$$x(t) = x^0 + \int_0^t (Vx)(s) \, ds, \quad t \in [0, T), \tag{18}$$

which can obviously be rewritten as

$$x(t) = (\mathcal{V}x)(t), \quad t \in [0, T), \tag{19}$$

with

$$(\mathcal{V}x)(t) = x^0 + \int_0^t (Vx)(s)\, ds, \quad t \in [0, T). \tag{20}$$

While (19) is formally identical to (16), we must also notice that formula (20) defines a causal operator, whenever V is a causal operator.

Therefore, equation (13) can be transformed (including the initial condition (3)) into a functional equation of the form (16), with causal operator.

The situation described above is usually encountered in the classical framework, when we transform the differential equation into an integral equation, and then apply various procedures of investigation.

The equation (10), usually termed an integrodifferential equation, also belongs to the class of functional differential equation described by (13), with the choice

$$(Vx)(t) = f\left(t, x(t), \int_0^t K(t, s, x(s))\, ds\right).$$

In subsequent chapters, we will have many opportunities to identify various functional equations as members of the wider class of functional or functional differential equations with causal operators.

In concluding this section, let us note that the integral operator in equation (9), which is not causal in general, reduces to the Volterra operator (12), if $K(t, s, x) \equiv 0$ for $t < s < T$. In many texts on integral equations, this remark is aimed at showing that (classical) Volterra operators are special cases of Fredholm operators. While this conclusion is formally true, it should be emphasized that the theory of classical integral equations of Volterra type has known, during the last decades, a tremendous development in the nonlinear case, which has not yet been matched by the theory of Fredholm-type equations. Of course, the linear case provided the elements of spectral theory (of operators) and gained a general acclaim within a short time after Fredholm published his celebrated paper. In the nonlinear case the study of Volterra-type equations constitutes a more advanced theory than that of Fredholm-type equations. Moreover, the Volterra equations with classical or abstract (causal) operators represent a more adequate tool in investigating evolutionary processes.

To substantiate the above considerations, we refer the reader to the following references: C. Corduneanu [10], G. Gripenberg, S.O. Londen and O. Staffans [1].

In concluding this section, we want to point out that for causal operators a notation (formally) identical to that encountered for general delay equations (see, for instance J.K. Hale [1]) can be used (see C. Corduneanu [10]). Namely, a representation of the form

$$(Vx)(t) = \mathcal{V}(t, x_t), \tag{21}$$

where, for each $t \in (0, T)$, $\mathcal{V}(t, y)$ is a functional on some function space $E = E([0, t], \mathbb{R}^n)$. The following example is illustrative: consider the classical operator V given by formula (17); then

$$\mathcal{V}(t, y_t) = f(t) + \int_0^t K(t, s, y(s))\, ds. \tag{22}$$

It is useful to note that $\mathcal{V}$ takes its values in $\mathbb{R}^n$, for each fixed t, while this family of functionals, $t \in [0, T)$, defines the operator from $E([0, T), \mathbb{R}^n)$ into itself.

1.3 Causal operators in applications

There are many opportunities to use causal operators in applied problems, and the engineering literature offers countless examples in this regard. Only part of the pertaining references are included in our list. See, for instance, W.J. Rugh [1] and M. Schetzen [1].

Let us formulate a simple problem in *optimal control theory*. Consider the functional to be minimized:

$$C(x, u) = \int_0^T F(t, x(t), u(t)) \, dt, \tag{23}$$

where the state variable x is related to the control u by the equation

$$\dot{x}(t) = (Ax)(t) + (Bu)(t), \tag{24}$$

with A and B causal operators on certain function spaces $E([0, T], \mathbb{R}^n)$, resp. $E_0([0, T], \mathbb{R}^m)$. An initial condition of the form (3) must be attached, in order to assure uniqueness of $x(t)$, for each given $u(t)$.

The problem one wants to solve is finding an *optimal control* $u^*(t)$, such that

$$C(x^*, u^*) = \min C(x, u), \tag{25}$$

where x^* is determined by (24) and (3).

Usually, some restrictions are imposed on the control function $u(t)$.

The above-formulated problem encompasses a wide variety of optimal control problems, including classical ones, when (24) is an ordinary differential equation, a delay equation or an integrodifferential equation of Volterra type.

A more detailed discussion of optimal control problems involving causal operators will be conducted in Chapter 7 (Applications).

Another example of the application of causal operators is provided by a problem in *linear continuous programming*. It originates with Bellman, and the case described below is a generalization of that problem.

The cost functional to be minimized is

$$C(x) = \int_0^T \langle c(t), x(t) \rangle \, dt, \tag{26}$$

where $\langle \cdot, \cdot \rangle$ denotes the scalar product in $\mathbb{R}^n$, $c(t)$ being assigned. There is a restriction on $x(t)$, namely

$$(Gx)(t) \leq b(t), \quad t \in [0, T], \tag{27}$$

with G a causal map from the underlying function space into $\mathbb{R}$, and $b(t)$ a given function. Other restrictions may be imposed on $x(t)$, such as $x(t) \geq 0$ on $[0, T]$ (i.e., each component of $x(t)$ is nonnegative).

It is appropriate to notice that a discretization procedure can be applied to find approximate solutions to (26) and (27).

For instance, if we want to minimize $C(x)$, under constraints (27) and $x(t) \geq 0$, the usual Euler method leads to the following problem of *linear programming*: for a given partition $0 = t_0 < t_1 < \cdots < t_n = T$ of the interval $[0, T]$, find

$$\min \sum_{k=1}^{n} (t_k - t_{k-1}) \langle c_k, x_k \rangle, \tag{28}$$

under the constraints

$$(G\tilde{x})(t_k) \leq b(t_k), \quad k = 0, 1, \ldots, n, \tag{29}$$

$$x_k \geq 0, \quad k = 1, 2, \ldots, n. \tag{30}$$

In (28), c_k stands for an intermediate value of $c(t)$, with t between t_{k-1} and t_k, while $\tilde{x}$ stands for the (continuous) function whose graph is the Euler polygon joining the points, $(t_k, x_k), k = 0, 1, 2, \ldots, n$. Obviously, the problem is to minimize a linear function within a given polyhedron (defined by (29), (30)). The classical procedure can be applied.

Usually, as in the original Bellman problem, (27) has the form

$$A(t)x(t) + \int_0^t B(t, s)x(s)\,ds \leq b(t), \tag{31}$$

where A and B stand for matrix-valued functions. For recent contributions to the subject and pertinent references, see Chapter 7.

The last example we will discuss in this section is related to the *dynamics of nuclear reactors*. An uncontrolled nuclear reactor is governed by the system of functional equations

$$\begin{cases} \dot{\rho}(t) = \sum_{k=1}^{m} \beta_k \Lambda^{-1}[\rho(t) - \eta_k(t)] - P\Lambda^{-1}[1 + \rho(t)]v(t), \\[2mm] \dot{\eta}_k(t) = \lambda_k[\rho(t) - \eta_k(t)], \quad k = 1, 2, \ldots, m, \\[2mm] v(t) = (\alpha\rho)(t). \end{cases} \tag{32}$$

The meaning of the variables involved in (32) is: ρ represents the reactor power, η_k are the flows of delayed neutrons, v is the reactivity feedback and is expressed as a causal functional which depends on ρ, while β_k, P, Λ and λ_k, $k = 1, 2, \ldots, m$, are constants. They have physical significance and details can be found in C. Corduneanu [10].

For us, in the present context, the important feature consists in the fact that the functional α in (32) is *causal* and it is not known analytically (in general). For instance, it can be chosen (see C. Corduneanu [10]) in the form

$$(\alpha\rho)(t) = \alpha_0\rho(t) + \sum_{k=1}^{\infty} \alpha_k \rho(t - t_k) + \int_{-\infty}^{t} \beta(t - s)\rho(s)\,ds,$$

where t_k are positive numbers and

$$\sum_{k=1}^{\infty} |\alpha_k| < \infty, \quad |\beta| \in L^1(0, \infty).$$

Let us also notice that substituting the second set of equations in (32) by

$$\eta_k(t) = \lambda_k \int_{-\infty}^{t} \exp\{-\lambda_k(t-s)\}\,ds, \quad k = 1, 2, \ldots, n,$$

and proceeding by elimination, one obtains a single equation instead of (32), namely

$$\dot{\rho}(t) = -\Lambda^{-1}\left(\sum_{k=1}^{m}\beta_k\right)\rho(t) + \Lambda^{-1}\int_{-\infty}^{t}\rho(s)\exp\{-\lambda_k(t-s)\}\,ds$$

$$- P\Lambda^{-1}[1 + \rho(t)](\alpha\rho)(t). \tag{33}$$

Equation (33) is a nonlinear causal functional equation; one may call it a functional inte-grodifferential equation, showing the complexity of the mathematical apparatus necessary to describe phenomena occuring in the modern technological world.

The investigation in depth of an equation like (33) is a matter of future consideration, if we want to preserve some generality for the causal functional α. In particular, stability (or asymptotic) stability of the zero solution of (33) is of primary interest in the physical context.

More discussions and examples on the application of causal operators will be included in Chapter 7 of this book.

2 Auxiliary concepts

2.1 Abstract spaces

Set theory was developed in the last part of the nineteenth century (Georg Cantor), and continued to flourish during the twentieth century. Set theory has provided the background for the development of new branches of modern mathematics. This theory is primarily due to M. Fréchet (1906), who introduced the concept of *metric space*. The definition of a metric space can be formulated as follows:

Let S be a set, and $d : S \times S \to \mathbb{R}_+ = [0, \infty)$ a map satisfying the conditions:

1   $d(x, y) = d(y, x)$;
2   $d(x, y) \geq 0$, the equality being possible only for $x = y$;
3   $d(x, y) \leq d(x, z) + d(z, y)$.

The couple (S, d) is called a *metric space* if conditions 1–3 are satisfied where x, y, z denote arbitrary elements of S.

The map d is called the *metric* of the space (S, d), or *distance* (between the elements of S). Sometimes the elements of S are called *points*.

The property 3, known as the *triangle* inequality, represents a generalization of the elementary geometry result expressed as: in a given triangle, the length of any side is smaller than the sum of the lengths of other two sides.

A very important consequence of the definition of a metric space is the fact that it allows one to extend the classical concept of *convergence*.

Let $\{x_n; \ n \geq 1\} \subset S$ be a sequence. Then $\{x_n\}$ is said to be *convergent* to the point $x \in S$, if

$$\lim d(x_n, x) = 0, \quad \text{as } n \to \infty. \tag{1}$$

One says that x is the limit of the sequence $\{x_n\}$.

Based on the triangle inequality for d, one can easily prove that for any convergent sequence, the limit is *unique*.

In other words, from (1) and

$$\lim d(x_n, \tilde{x}) = 0, \quad \text{as } n \to \infty, \tag{2}$$

one must obtain $x = \tilde{x}$. Indeed, one can write

$$d(x, \tilde{x}) \leq d(x, x_n) + d(x_n, \tilde{x})$$

for any $n \geq 1$, which, together with (1) and (2), lead to $d(x, \tilde{x}) = 0$. The property 2 of d now applies, and yields $x = \tilde{x}$.

Let $\{x_n\} \subset S$ be a convergent sequence. Then the *Cauchy property* (criterion) is valid: for each $\varepsilon > 0$, there exists a natural number $N = N(\varepsilon)$, such that

$$(n, m \geq N(\varepsilon)) \Longrightarrow d(x_n, x_m) < \varepsilon. \tag{3}$$

Indeed, according to the definition, for each $\varepsilon > 0$ there exists $N = N(\varepsilon)$, such that

$$d(x_n, x) < \frac{\varepsilon}{2} \quad \text{for } n \geq N(\varepsilon), \tag{4}$$

where $x = \lim x_n$ as $n \to \infty$. Hence

$$d(x_n, x_m) \leq d(x_n, x) + d(x, x_m) < \frac{\varepsilon}{2} + \frac{\varepsilon}{2} = \varepsilon,$$

provided $n, m \geq N(\varepsilon)$, which proves (3).

The converse of the Cauchy property is not necessarily true in any metric space.

A metric space is called *complete* if any sequence satisfying the Cauchy property is convergent.

As we know from real number theory, the set $\mathbb{R}$ of reals, with the usual distance $d(x, y) = |x - y|$ is a complete metric space.

If $M \subset S$ is a subset, then $x \in M$ is called a *limit point* of M if there exists $\{x_n\} \subset M$, such that $\{x_n\}$ converges to x. The set of limit points of S is called the *closure* of M and is denoted by $\overline{M}$.

A subset $M \subset S$ is called *closed* if $M = \overline{M}$.

It is easy to prove that a closed subset $M \subset S$ is a complete metric space (with the same distance d as in S), anytime S is complete.

A metric space (S, d) is called *compact* if any sequence $\{x_n\} \subset S$ has a convergent subsequence. In other words, there exists a sequence $\{n_k;\ k \geq 1\}$ of integers, such that

$$\lim x_{n_k} \text{ exists as } k \to \infty.$$

Sometimes, instead of compact one may find the term *sequentially compact*.

A set $C \subset S$ is called *compact* if it is compact in the sense of the above definition when itself regarded as a metric space (C, d).

It can be easily proven that a compact set $C \subset S$ is necessarily closed. This prompts the next definition.

A set $C \subset S$ is called *relatively compact* if its closure $\overline{C}$ is compact.

A classical result about real numbers (or the real line), known as the Bolzano–Weierstrass theorem, states that any bounded subset of R is realtively compact. In the original formulation, this property was every infinite bounded set in $\mathbb{R}$ contains a convergent subsequence.

The following property can be easily proven: if (S, d) is a compact metric space, then it is complete.

Indeed, if one takes a Cauchy sequence $\{x_n\} \subset S$, it must contain a convergent subsequence $\{x_{n_k}\}$. The limit of this subsequence is actually the limit of $\{x_n\}$.

A compact space (S, d) always has a *finite diameter*:

$$D(S) = \sup\{d(x, y);\ x, y \in S\}. \tag{5}$$

Since $d(x, y)$ is a continuous map from $S \times S$ into $\mathbb{R}_+$ (because one always has $|d(x_n, y_n) - d(x, y)| \leq d(x_n, x) + d(y, y_n)$), the proof can be conducted along the same lines as in the case of the classical Weierstrass theorem on the boundedness of continuous functions on compact sets.

It is interesting to note that for any metric space (S, d), there exists a metric

$$\tilde{d}(x, y) = \frac{d(x, y)}{1 + d(x, y)}, \quad x, y \in S, \tag{6}$$

such that $(S, \tilde{d})$ has (obviously) bounded diameter. At the same time, we should notice the fact that $\tilde{d}(x_n, x) \to 0$ as $n \to \infty$ iff $d(x_n, x) \to 0$ as $n \to \infty$. In other words, convergence is the same in (S, d) as in $(S, \tilde{d})$. One says that the two metrics are *equivalent*.

We will close these generalities about metric spaces with a *compactness* criterion.

The set $M \subset S$ is *relatively compact* iff: for every $\varepsilon > 0$, there exists a finite ε-net for M.

This means that, given $\varepsilon > 0$, one can find some points $x_k \in M, k = 1, 2, \ldots, n = n(\varepsilon)$, such that for every $x \in M$, there exists $j, 1 \leq j \leq n$, with $d(x_j, x) < \varepsilon$.

For a proof of the above criterion we refer the reader to A. Friedman [1].

Among the complete metric spaces, we shall deal now with two remarkable classes, with numerous applications in the theory of functional equations: namely, Banach spaces and Hilbert spaces. In the next section, we shall discuss several types of function spaces (which are of primary interest in the theory of functional equations), all of them being either Banach or Hilbert spaces.

Before formulating the definitions of Banach spaces or Hilbert spaces, it is appropriate to recall the definition of a *linear space* (or *vector space*). Besides the metric structure that Banach/Hilbert spaces possess, they are also organized algebraically as linear spaces.

Let X be a set of elements, and $\mathbb{R}$ denote the real line. Assume two mappings are defined as follows: the first is from $X \times X$ into X, say $(x, y) \to x + y$ and is called *addition*, while the second is from $R \times X$ into X, say $(\lambda, x) \to \lambda x$, which is called *scalar multiplication*. Moreover, the following axioms are assumed:

1 X is an Abelian group with respect to addition;
2 scalar multiplication is associative, i.e., $\lambda(\mu x) = (\lambda\mu)x$, for any $\lambda, \mu \in R$ and $x \in X$;
3 the distributivity of both addition and scalar multiplication, i.e., $(\lambda + \mu)x = \lambda x + \mu x$
    and $\lambda(x + y) = \lambda x + \lambda y$, for any $\lambda, \mu \in R$, and $x, y \in X$;
4 $1x = x$, for any $x \in X$.

If X satisfies all conditions mentioned in the preceding paragraph, then X is called a *linear space* (over the field of reals).

A more general concept of linear space is obtained if the field $\mathbb{R}$ is replaced by another field (of scalars), say $\mathcal{F}$. In particular, one can choose $\mathcal{F} = \mathcal{C} =$ the complex field.

The term *Abelian group* means that the operation of addition is commutative, associative, admits a null element θ, such that $x + \theta = x$ for any $x \in X$, and for each $x \in X$ there exists $-x \in X$, such that $x + (-x) = \theta$.

We can now define the concept of a *normed linear space*, which is somewhat more general that the concept of a Banach space.

Let X be a linear space, as defined above. A *norm* on X is a map from X into $\mathbb{R}_+$, say $x \to \|x\|$, satisfying the following conditions:

1 $\|x\| \geq 0$ and $\|x\| = 0$ only if $x = \theta$;
2 $\|\lambda x\| = |\lambda| \|x\|$;
3 $\|x + y\| \leq \|x\| + \|y\|$, for all $x, y \in X$.

Then X is called a *linear normed space*.

The following proposition is almost obvious: If X is a linear normed space, then X is a metric space, with the metric $d(x, y) = \|x - y\|$.

We are now able to introduce the concept of a Banach space (which we will denote by B).

A *Banach space* B is a linear normed space, which is complete in the metric $d(x, y) = \|x - y\|$.

An example of a Banach space is the Euclidean n-space $\mathbb{R}^n$.

In other words, if $x = (x_1, x_2, \ldots, x_n)$, $y = (y_1, y_2, \ldots, y_n) \in \mathbb{R}^n$, then $x + y = (x_1 + y_1, \ldots, x_n + y_n)$, and $\lambda x = (\lambda x_1, \ldots, \lambda x_n)$. We ask the reader to check the conditions in the definition of a norm, for each of the candidates: $\|x\| = \max(|x_i|; 1 \leq i \leq n)$; $\|x\| = |x_1| + |x_2| + \cdots + |x_n|$; $\|x\| = (x_1^2 + x_2^2 + \cdots + x_n^2)^{1/2}$.

More sophisticated examples can be constructed. For instance, if the elements of B are bounded sequences of real numbers, $x = (x_1, x_2, \ldots, x_k, \ldots)$, $|x_k| \leq M_x$, $k = 1, 2, \ldots$, then V is a Banach space if addition and scalar multiplication are defined by $x + y = (x_1 + y_1, \ldots, x_k + y_k, \ldots)$, and $\lambda x = (\lambda x, \ldots)$. A norm can be defined by $\|x\| = \sup(|x_k|; k = 1, 2, \ldots)$.

While all properties of the norm in its definition can be easily proven, the completeness of the space requires more serious effort. Nonetheless, the key resides in the classical (but elementary) Cauchy criterion of convergence on the real line.

A *Hilbert space* is a Banach space in which the norm is derived from an *inner product* (or *scalar product*).

Let us define the concept of *inner product*. Assume X is a linear space, as defined above. We say that an inner product is defined on X, if a mapping from $X \times X$ into $\mathbb{R}$ is given, say $(x, y) \rightarrow \langle x, y \rangle$, such that:

1 $\langle x, x \rangle \geq 0$ for any $x \in X$, and $\langle x, x \rangle = 0$ only for $x = \theta$;
2 $\langle x, y \rangle = \langle y, x \rangle$ for any $x, y \in X$;
3 $\langle x + y, z \rangle = \langle x, z \rangle + \langle y, z \rangle$;
4 $\langle \lambda x, y \rangle = \lambda \langle x, y \rangle$, for any $x, y \in X$ and $\lambda \in \mathbb{R}$.

In many applications, instead of $\mathbb{R}$ one deals with $\mathbb{C}$. Then, instead of the symmetry condition one chooses the property $\langle x, y \rangle = \overline{\langle y, x \rangle}$, the bar meaning the conjugate complex number.

It can be easily seen that any inner product on X generates a norm, by choosing $\|x\| = \langle x, x \rangle^{1/2}$. The triangle inequality for the norm is obtained by establishing first the Schwarz inequality $|\langle x, y \rangle| \leq \|x\| \|y\|$. This follows from $\langle x + \lambda y, x + \lambda y \rangle \geq 0$ for any $x, y \in X$ and $\lambda \in R$. It means that $\|x\|^2 + 2\lambda \langle x, y \rangle + \lambda^2 \langle y, y \rangle \geq 0$ for all λ. This is obviously possible only in case $\langle x, y \rangle^2 \leq \langle x, x \rangle \cdot \langle y, y \rangle$, which shows the validity of the Schwarz inequality for the inner product. The triangle inequality for the norm $\|x\| = \langle x, x \rangle^{1/2}$ is obtained from $\|x + y\|^2 = \|x\|^2 + 2\langle x, y \rangle + \|y\|^2$, taking into account the Schwarz inequality.

Since any inner product induces a norm on a linear space, it follows that in order to have a *Hilbert space*, one has to start with a linear space, endowed with an inner product, such that this space is *complete* with respect to the norm $\|x\| = \langle x, x \rangle^{1/2}$. One usually denotes a Hilbert space by H.

Historically, Hilbert spaces were discovered prior to Banach spaces. The first example of a Hilbert space, defined by Hilbert himself, is the space $\ell^2(\mathbb{R})$ of sequences $x = (x_1, x_2, \ldots, x_k, \ldots)$, such that

$$\sum_{k=1}^{\infty} x_k^2 < +\infty, \tag{7}$$

with the usual addition and scalar multiplication, and the inner product

$$\langle x, y \rangle = \sum_{k=1}^{\infty} x_k y_k. \tag{8}$$

The series in (8) is always absolutely convergent for $x, y \in \ell^2$, because $2|x_k y_k| \le x_k^2 + y_k^2$. If we deal with the Hilbert space $\ell^2(\mathbb{C})$, of complex sequences, then (7) must be replaced by $\sum_{k=1}^{\infty} |x_k|^2 < \infty$, while (8) must be $\langle x, y \rangle = \sum_{k=1}^{\infty} x_k \bar{y}_k$.

As expected, the only property presenting some difficulty in being checked is *completeness*. Assume that $\{x^{(k)}\} = \{(x_1^{(k)}, x_2^{(k)}, \dots, x_p^{(k)}, \dots)\} \subset \ell^2$ is a Cauchy sequence:

$$\|x^{(k)} - x^{(m)}\| < \varepsilon, \quad \text{for } k, m \ge N(\varepsilon).$$

Since

$$|x_p^{(k)} - x_p^{(m)}| \le \|x^{(k)} - x^{(m)}\| < \varepsilon, \quad \text{for } k, m \ge N(\varepsilon), \tag{9}$$

it follows that, for each fixed p, $\{x_p^{(k)}\}$ is a Cauchy sequence in $\mathbb{R}$. Therefore, it is convergent. Let $\lim x_p^{(k)} = x_p$, as $k \to \infty$, be its limit. Then $x = (x_1, x_2, \dots, x_p, \dots) \in \ell^2$ and $x^{(k)} \to x$ as $k \to \infty$.

Indeed, since $x = (x - x^{(k)}) + x^{(k)}$, it suffices to show that $x - x^{(k)} \in \ell^2$ for at least one k. But (9) implies

$$\sum_{j=1}^{p} \left(x_j^{(k)} - x_j^{(m)} \right)^2 < \varepsilon^2, \quad k, m \in N(\varepsilon), \tag{10}$$

for any natural number p. From (10) we derive

$$\sum_{j=1}^{p} \left(x_j^{(k)} - x_j \right)^2 \le \varepsilon^2, \quad k \ge N(\varepsilon) \tag{11}$$

if we let $m \to \infty$. Now, since p is arbitrary in (11), one obtain

$$\sum_{j=1}^{\infty} \left(x_j^{(k)} - x_j \right)^2 \le \varepsilon^2, \tag{12}$$

which shows that $x^{(k)} - x \in \ell^2$, if $k \ge N(\varepsilon)$.

From (11) we also see that $\|x^{(k)} - x\| \le \varepsilon$ for $k \ge N(\varepsilon)$. This means that $x^{(k)} \to x$ in ℓ^2, which ends the proof of completeness.

In Hilbert spaces, by means of the inner product, one can proceed to the construction of the (Euclidean) geometry in infinite dimension. Indeed, since $\|x\|$ is the length of the vector x, the cosine of the angle α formed by the vectors x and y is determined from the formula $\langle x, y \rangle = \|x\| \|y\| \cos \alpha$. The area of the triangle formed by the vectors x, y and $x - y$ is $\frac{1}{2}\|x\| \|y\| \sin \alpha$, etc.

The concept of orthogonality of vectors in a Hilbert space, defined by $x \perp y$ iff $\langle x, y \rangle = 0$, plays an important role in theory and applications.

The following question is natural in this context: When is a Banach space a Hilbert space? In other words, for what category of Banach spaces can we construct an inner product that generates the given norm?

The answer to this question is very simple. The only condition the norm of a Banach space B must satisfy in order to be derived from an inner product is

$$\|x + y\|^2 + \|x - y\|^2 = 2 \left(\|x\|^2 + \|y\|^2 \right), \tag{13}$$

for any $x, y \in B$.

The equality (13) expresses the classical theorem in elementary geometry that states: in any parallelogram, the sum of the squares of the diagonals is equal to the sum of the squares of its sides.

If the norm of the Banach space B satisfies (13), then the inner product can be defined by the formula

$$\langle x, y \rangle = \tfrac{1}{4} \left(\|x + y\|^2 - \|x - y\|^2 \right). \tag{14}$$

In concluding this section, we shall dwell upon a concept of metric space, usually known as *Fréchet space* (even though the general concept of a metric space is due to Fréchet).

Let X be a linear space, endowed with a distance function (metric) d satisfying the condition

$$d(x + z, y + z) = d(x, y), \tag{15}$$

for any $x, y, z \in X$. This property means the *invariance* of the distance with respect to translations.

If in (15) one chooses $z = -y$, then the following formula is obtained:

$$d(x, y) = d(x - y, \theta). \tag{16}$$

In particular, if X is a normed space, the formula (16) becomes $d(x, y) = \|x - y\|$, which is the usual one.

A *Fréchet space* is a linear space endowed with an invariant metric with respect to translations and is complete with respect to this metric.

It is usual to define a Fréchet space by means of the use of seminorms.

A *seminorm* on the linear space X is a map from X into $\mathbb{R}_+$, say $x \to |x|$, such that:

1    $|x| \geq 0$;
2    $|\lambda x| = |\lambda||x|$;
3    $|x + y| \leq |x| + |y|$.

The only difference with respect to a norm consists in the fact that $|x| = 0$ does not necessarily imply $x = \theta$. Based on properties 2 and 3, one sees that the set of $x \in X$ for which $|x| = 0$ is a linear manifold in X.

If $\{|x|_k; \ k \geq 1\}$ is a family of seminorms on X, one says that this family is *sufficient* iff from $|x|_k = 0, k \geq 1$, one derives $x = \theta$.

Now assume X is a linear space endowed with a (countable) sufficient family of seminorms, say $\{|x_k|; \ k \geq 1\}$. Then X is a metric space, with the distance function

$$d(x, y) = \sum_{k=1}^{\infty} \frac{1}{2^k} \frac{|x - y|_k}{1 + |x - y|_k}. \tag{17}$$

Checking the properties of the metric is an elementary exercise (left to the reader), with the possible exception of the triangle inequality. For the latter, one must rely on the elementary inequality

$$\frac{|a+b|}{1+|a+b|} \leq \frac{|a|}{1+|a|} + \frac{|b|}{1+|b|}, \tag{18}$$

in which $a, b \in \mathbb{R}$.

If X is complete with respect to the metric defined by (17), then it is a *Fréchet space*.

From (17) it is obvious that the metric $d(x, y)$ is translation invariant.

Let us mention that this way of defining Fréchet spaces shall be used in the next section in dealing with function spaces.

For a more detailed discussion of the concepts considered in this section, we refer the reader to the excellent, and very concise, books by A. Friedman [1] or I. Gohberg and S. Goldberg [1].

2.2 Function spaces

The study of various classes of functional equations requires the use of *function spaces*. These spaces are usually Banach spaces or Hilbert spaces. In some cases, it will be recommended to use *Fréchet spaces*, as defined in Section 2.1. We shall deal with two categories of function spaces: spaces of *continuous* functions and spaces of *measurable* functions (more precisely, locally integrable in the Lebesgue sense). These spaces will provide the framework for the investigation of basic problems related to various classes of functional equations, as those mentioned in Chapter 1.

Let us start with the space of all continuous maps from a closed interval $[t_0, T]$, into the space $\mathbb{R}^n$. The complete definition of this space, denoted by $C([t_0, T], \mathbb{R}^n)$, is as follows: in the *linear space* of all continuous maps from $[t_0, T]$ into $\mathbb{R}^n$, one defines the *norm*

$$\|x\| = \sup\{|x(t)|; \quad t \in [t_0, T]\}, \tag{19}$$

where $|\cdot|$ denotes the Euclidean norm in $\mathbb{R}^n$ (i.e., $|x| = (x_1^2 + x_2^2 + \cdots + x_n^2)^{1/2}$).

It is easy to check the conditions specific to a norm. So far, we have introduced a linear *normed* space.

The completeness of this space will allow us to conclude that $C([t_0, T], \mathbb{R}^n)$ is a *Banach space*. Indeed, let us notice that convergence with respect to the norm $\|\cdot\|$, defined by (19), is the *uniform convergence* of continuous functions on the interval $[t_0, T]$. If $x_n(t) \to x(t)$ in the sense of norm (19), usually called the *supremum norm*, then we can write for each $\varepsilon > 0$

$$\|x_n(t) - x(t)\| < \varepsilon, \quad \text{for } n \geq N(\varepsilon), \tag{20}$$

which implies, for all $t \in [t_0, T]$,

$$|x_n(t) - x(t)| < \varepsilon, \quad \text{for } n \geq N(\varepsilon). \tag{21}$$

The formula (21) represents exactly the condition of uniform convergence on $[t_0, T]$, of $\{x_n(t)\}$ to $x(t)$. As we know from classical calculus, the function $x(t)$ is continuous on $[t_0, T]$. This suffices to conclude the completeness of $C([t_0, T], \mathbb{R}^n)$.

If one starts with a Cauchy sequence in the normed space defined above, the conclusion reached will be the same.

The Banach space $C([t_0, T], \mathbb{R}^n)$ is one of the most often used in the theory of functional equations. That's why, knowing a compactness criterion in $C([t_0, T], \mathbb{R}^n)$ is of great importance.

The following result is known as the *Arzelà–Ascoli criterion*.

Let $M \subset C([t_0, T], \mathbb{R}^n)$. Then, M is relatively compact with respect to uniform convergence on $[t_0, T]$, iff:

a M is *bounded* in C, i.e., there exists a positive number $A > 0$, such that

$$|x(t)| \leq A, t \in [t_0, T], \quad \text{for each } x \in M; \tag{22}$$

b M is *equicontinuous*, which means that for any $\varepsilon > 0$, there exists $\delta = \delta(\varepsilon) > 0$, such that for $t, s \in [t_0, T]$,

$$|x(t) - x(s)| < \varepsilon, \quad \text{for } |t - s| < \delta, \tag{23}$$

for every $x \in M$.

The proof of this criterion can be found in many books. See, for instance, A. Friedman [1] or C. Corduneanu [10].

One simple circumstance in which (23) takes place is when all functions in M satisfy the Lipschitz condition

$$|x(t) - x(s)| \leq L|t - s|, \tag{24}$$

on $[t_0, T]$, with same constant $L > 0$.

Sometimes, the supremum norm defined by (19) should be replaced by an equivalent norm (i.e., inducing uniform convergence of function sequences in C). One simple way of constructing such norms is the introduction of a *weighted norm*.

Let $g(t)$ be continuous and positive on $[t_0, T]$. Then, defining

$$\|x\|_g = \sup\{g(t)|x(t)|; \ t \in [t_0, T]\} \tag{25}$$

leads to an equivalent norm to (19), on C.

Indeed, if we denote $g_m = \inf g(t), g_M = \sup g(t), t \in [t_0, T]$, these numbers are positive and we have the inequalities

$$g_m|x| \leq |x|_g \leq g_M|x|, \quad x \in C. \tag{26}$$

The use of a weighted norm could bring some simplifications in subsequent chapters, and leads to better results.

On the other hand, the type of convergence induced by the norm $\|x\|_g$ being the same as in the case of norm (19), the Arzelà–Ascoli compactness criterion keeps its valadity, regardless of the particular norm we are using (supremum or weighted).

The situation encountered above changes significantly if instead of continuous functions on $[t_0, T]$, we deal with continuous functions on the semi-open interval $[t_0, T)$, the possibility $T = +\infty$ not being excluded. Since a continuous function on $[t_0, T)$ may not be bounded, the formula (19) may not make sense.

Actually, the space $C([t_0, T), \mathbb{R}^n)$, which we shall soon define, is not a normed space. It can be organized as a Fréchet space. Namely, let us consider a sequence of real numbers

$\{t_k;\ t \geq 1\} \subset (t_0, T)$, such that $t_k \uparrow T$ as $k \to \infty$. Then define, for any continuous function on $[t_0, T)$, with values in $\mathbb{R}^n$, the quantities

$$|x|_m = \sup\{|x(t)|;\ t \in [t_0, t_m]\},\quad m \geq 1. \tag{27}$$

It is obvious that $x \to |x|_m$ is a seminorm on the linear space of all maps from $[t_0, T)$ into $\mathbb{R}^n$. Moreover, $x = \theta$ iff $|x|_m = 0, m \geq 1$. Hence, this family of seminorms is *sufficient*. Therefore, as seen in Section 2.1, the function

$$d(x, y) = \sum_{m=1}^{\infty} \frac{1}{2^m} \frac{|x - y|_m}{1 + |x - y|_m} \tag{28}$$

is a metric on $C([t_0, T), \mathbb{R}^n)$, which is translation invariant. Hence, it remains to show that $C([t_0, T), \mathbb{R}^n)$ is complete with respect to this metric, in order to conclude that it is a Fréchet space.

It is now appropriate to see what kind of convergence is defined by the metric (28). Since for any fixed $m \geq 1$ one obtains from (28)

$$2^m d(x, y) \geq |x - y|_m (1 + |x - y|_m)^{-1},$$

it means that convergence with respect to the metric $d(x, y)$ implies uniform convergence on each interval $[t_0, t_m]$. Moreover, since any compact interval belonging to $[t_0, T)$ is contained in some interval $[t_0, t_m]$, this means that convergence in Fréchet space is uniform convergence on each compact interval (or set) in $[t_0, T)$. Vice versa, if a sequence of functions in $C([t_0, T), \mathbb{R}^n)$ is uniformly convergent on each compact set in $[t_0, T)$, which is the same thing as uniform convergence on each $[t_0, t_m], m \geq 1$, then this sequence converges in the sense of the metric $d(x, y)$ given by (28).

Indeed, assume $\{x^{(k)}\} \subset C([t_0, T), \mathbb{R}^n)$ converges uniformly to x on each interval $[t_0, t_m], m \geq 1$. We need to show that, given $\varepsilon > 0$, one can find $N(\varepsilon)$ with the property

$$d(x^{(k)}, x) < \varepsilon,\quad \text{for } k \geq N(\varepsilon). \tag{29}$$

Since each fraction in the right-hand side of (28) is less than 1, the remainder in the series there is dominated by $\sum_{m=N+1}^{\infty} 2^{-m} = 2^{-N}$. Hence, the remainder can be made arbitrarily small, provided N is sufficiently large (for instance, $2^{-N} < \varepsilon/2$ if $N > \ln(2/\varepsilon)/\ln 2$). Therefore, one can always write

$$d(x^{(k)}, x) < \sum_{m=1}^{N} 2^{-m} |x^{(k)} - x|_m \left(1 + |x^{(k)} - x|_m\right)^{-1} + \frac{\varepsilon}{2} \tag{30}$$

for $N > \ln(2/\varepsilon)/\ln 2$. If we fix such an N, then we can obviously choose k large enough, such that

$$\sum_{m=1}^{N} 2^{-m} |x^{(k)} - x|_m \left(1 + |x^{(k)} - x|_m\right)^{-1} < \frac{\varepsilon}{2}, \tag{31}$$

because of the uniform convergence of $\{x^{(k)}\}$ on any $[t_0, t_m]$. Actually, in this case we have to look at the interval $[t_0, t_N]$.

Combining (30) and (31) we obtain (29) for sufficiently large $N(\varepsilon)$, which proves that the metric (28) defines uniform convergence on any compact interval of $[t_0, T)$.

It is also interesting to understand what compactness means in the Fréchet space $C([t_0, T), \mathbb{R}^n)$. Since we have established the fact that convergence in $C([t_0, T), \mathbb{R}^n)$ means uniform convergence on each compact interval in $[t_0, T)$, it follows that relative *compactness* in $C([t_0, T), \mathbb{R}^n)$ can be defined as follows: The set $M \subset C([t_0, T], \mathbb{R}^n)$ is relatively compact if from any sequence $\{x^{(k)}; \ k \geq 1\} \subset M$ one can extract a subsequence, say $\{x^{(k)}; \ k \geq 1\}$, which is uniformly convergent on any compact interval in $[t_0, T)$. If one considers now the restictions of the terms of the sequence $\{x^{(k)}\}$ to some interval $[t_0, t_m]$, where $\{t_m\}$ is as described above, then we realize that these restrictions must satisfy the conditions of the Arzelà–Ascoli criterion. This leads to the following criterion of compactness in $C([t_0, T), \mathbb{R}^n)$.

The set $M \subset C([t_0, T), \mathbb{R}^n)$ is relatively compact, iff it is bounded and equicontinuous on each compact interval in $[t_0, T)$.

The space $C([t_0, T), \mathbb{R}^n)$ with the metric (28) contains *subspaces* which are Banach spaces. This remark may be useful in discussing various kinds of asymptotic behavior at infinity ($T = +\infty$). One simple method of constructing such spaces is by using some weighted norms.

If $g : [t_0, \infty) \to \mathbb{R}_+$ is a continuous map, then we can consider the set of all maps in $C([t_0, \infty), \mathbb{R}^n)$ satisfying

$$\sup (|x(t)|/g(t); \ t \in [t_0, \infty)) < +\infty. \tag{32}$$

If one denotes by $C_g([t_0, \infty), \mathbb{R}^n)$ the set of all $x \in C$ satisfying (32), and by $\|x\|_g$ the quantity in the left-hand side of (32), one can easily see that C_g is a Banach space with the norm $\| \cdot \|_g$.

More details about the spaces $C_g([t_0, \infty], \mathbb{R}^n)$ can be found in C. Corduneanu [10], [23]. It is also possible to substitute for the scalar weight g a matrix weight.

Another space of continuous functions that will intervene in subsequent chapters is the space of continuously differentiable functions. The space $C^{(1)}([t_0, T], \mathbb{R}^n)$ is defined as follows:

Consider all maps from $[t_0, T]$ into $\mathbb{R}^n$, continuous together with their (first-order) derivatives. We understand that at t_0 the map has only the right derivative while at T it has only the left derivative.

Based on elementary properties of functions and their derivatives we see that the set considered above is a linear space (over reals). The usual norm for this space is given by

$$|x|_1 = \sup(|x(t)| + |\dot{x}(t)|; \ t \in [t_0, T]). \tag{33}$$

With this norm, the linear space becomes a normed space, and it is an elementary exercise to establish that it is *complete*. In other words, $C^{(1)}([t_0, T], \mathbb{R}^n)$ with norm (33) is a *Banach space*.

It is obvious that $C^{(1)}([t_0, T), \mathbb{R}^n) \subset C([t_0, T], \mathbb{R}^n)$. *Convergence* in $C^{(1)}$ means uniform convergence on $[t_0, T]$ for both functions and (first-order) derivatives.

Compactness is also reduced to the Arzelà–Ascoli criterion. Its conditions must be satisfied by both sets $\{x(t)\}$ and $\{\dot{x}(t)\}$.

The spaces $C^{(k)}([t_0, T], \mathbb{R}^n)$, $k \geq 2$, can be defined in a similar way.

We shall consider now function spaces whose elements are *measurable functions* of the real variable t. We begin with the Hilbert space $L^2([t_0, T], \mathbb{R}^n)$ of square integrable maps from the interval $[t_0, T]$, into the space $\mathbb{R}^n$. The condition defining the space $L^2([t_0, T], \mathbb{R}^n)$ is

$$\int_{t_0}^{T} |x(t)|^2 \, dt < +\infty, \tag{34}$$

where the integral is understood in the sense of Lebesgue. For the definition and basic properties of the Lebesgue's integral we invite the reader to consult the book by A. Friedman [1].

Usually, the norm $|\cdot|$ involved in (31) will be the Euclidean norm (i.e., $|x|^2 = |x_1|^2 + |x_2|^2 + \cdots + |x_n|^2$). Since $L^2([t_0, T], \mathbb{R}^n)$ is a linear space (to see that the sum of two functions in L^2 belongs to this space one uses the elementary inequality $|x + y|^2 \leq 2(|x|^2 + |y|^2)$, in order to prove that it can be organized as a Hilbert space it suffices to define the inner (scalar) product. It is easily seen that

$$\langle x, y \rangle = \int_{t_0}^{T} x(t) \cdot y(t) \, dt, \tag{35}$$

with $x \cdot y = x_1 y_1 + x_2 y_2 + \cdots + x_n y_n$, satisfies the conditions required for inner products. In particular,

$$\|x\|^2 = \int_{t_0}^{T} |x(t)|^2 \, dt, \tag{36}$$

and $\|x\| = 0$ is possible in this case, and only in the case $x(t) = 0$ almost everywhere (a.e.) on $[t_0, T]$.

The most delicate (and difficult) point in proving that $L^2([t_0, T], \mathbb{R}^n)$, in which the scalar product is given by (35), is a Hilbert space is represented by the proof of completeness of this space with respect to the norm given by (36). This proof consists of what is known in integration theory as the Riesz–Fischer theorem. Again, due to the technicality of this proof, we refer the reader to general references, such as A. Friedman [1] (the case L^p, $1 \leq p \leq \infty$ is actually dealt with).

The definition of $L^p([t_0, T], R^n)$, $1 \leq p < \infty$, is similar to the case $p = 2$. Namely, the condition

$$\int_{t_0}^{T} |x(t)|^p \, dt < \infty \tag{37}$$

means that the measurable function $x(t)$ belongs to L^p. For the case $p = \infty$, (37) is replaced by

$$\text{ess–sup} |x(t)| < \infty, \quad t \in [t_0, T). \tag{38}$$

The norm in L^p, $1 \leq p < \infty$, is defined as

$$|x|_p = \left\{ \int_{t_0}^{T} |x(t)|^p \, dt \right\}^{1/p}, \tag{39}$$

while for $p = \infty$ it is

$$\|x\|_\infty = \text{ess–sup} |x(t)|, \quad t \in [t_0, T]. \tag{40}$$

While for $p = 1$ and $p = \infty$ the properties of the norm are obvious, the case $1 < p < \infty$ requires a more technical approach and we shall not go into details here (see, for instance A. Friedman [1]). The case $p = 2$ is a consequence of the fact that L^2 is a Hilbert space. More precisely, the triangle inequality for the norm ($\|x + y\|_p \leq \|x\|_p + \|y\|_p$, x, $y \in L^p$) poses some difficulty, and it can be obtained by first establishing Hölder's inequality

$$\|xy\|_1 \leq |x|_p |y|_q, \tag{41}$$

for scalar x and y, where p and q are conjugate indices: $p^{-1} + q^{-1} = 1$. The cases $p = 1$, $q = \infty$ or $p = \infty$, $q = 1$ are allowed.

The completeness of L^p, $1 \leq p \leq \infty$, is basically a consequence of the Riesz–Fischer theorem whose proof can be found in the above-quoted reference, as well as in many textbooks on measure theory or integration.

In subsequent chapters we shall use the spaces $L^p([t_0, T], \mathbb{R}^n)$, $1 \leq p \leq \infty$, described above, as well as the spaces $L^p_{\text{loc}}([t_0, T), \mathbb{R}^n)$. In the engineering literature these spaces are oftenly called *extended spaces*.

The definition of $L^p_{\text{loc}}([t_0, T), \mathbb{R}^n)$, which is only a Fréchet space, can be easily formulated if we use the concept of seminorm. Let $t_n \uparrow T$ be an increasing sequence (the final result does not depend on the particular sequence we choose), and denote

$$|x|_p^{(m)} = \left\{ \int_{t_0}^{t_m} |x(t)|^p \, dt \right\}^{1/p}, \qquad m \geq 1, \tag{42}$$

which we assume to be *finite* for every $m \geq 1$. For $p = \infty$ one uses the corresponding norm of L^∞. By definition, $x \in L^p_{\text{loc}}([t_0, T), \mathbb{R}^n)$ iff it is measurable and all quantities in the right hand side of (42) are finite. The metric of L^p_{loc} is then defined as shown above, by using the seminorms (42):

$$d_p(x, y) = \sum_{m=1}^{\infty} \frac{1}{2^m} \frac{|x - y|_p^{(m)}}{1 + |x - y|_p^{(m)}}.$$

This metric induces L^p-convergence on any $[t_0, t_m] \subset [t_0, T]$, or – which is the same – L^p-convergence on any compact interval in $[t_0, T)$. In other words, $x^{(k)} \to x$ in $L^p_{\text{loc}}([t_0, T), \mathbb{R}^n)$, iff $|x^{(k)} - x|_p^{(m)} \to 0$ as $k \to \infty$, for any $m \geq 1$.

Compactness in L^p will also be necessary in subsequent chapters. Criteria for compactness are known in the literature and we shall state some of them here.

Riesz' criterion. Let $M \subset L^p([t_0, T], \mathbb{R}^n)$, $1 < p < \infty$. Necessary and sufficient conditions for the relative compactness of M are:

1 M is bounded in L^p;
2 $\int_{t_0}^{T} |x(t + h) - x(t)|^p \, dt \to 0$ as $h \to 0$, uniformly with respect to $x \in M$.

Remark   If $t + h \notin [t_0, T]$, then we extend $x(t)$ outside $[t_0, T]$ by letting $x(t) = \theta$.

Kolmogorov's criterion. Let $M \subset L^p([t_0, T], \mathbb{R}^n)$, $1 \le p < \infty$. Necessary and sufficient conditions for the relative compactness of M are:

1   M is bounded in L^p;
2   $x_h \to x$ as $h \to 0$, uniformly with respect to $x \in M$, where

$$x_h(t) = h^{-1} \int_t^{t+h} x(s)\, ds, \quad h > 0.$$

For the proof of the above compactness criteria, as well as other criteria, see L.V. Kantorovich and G.P. Akilov [1] or N. Dunford and J.T. Schwarz [1]. In a similar manner as in case of spaces $C([t_0, T], \mathbb{R}^n)$ and $C([t_0, T), \mathbb{R}^n)$, one can reduce the compactness in L^p_{loc} to the L^p-compactness on each compact interval in $[t_0, T)$.

Besides the above defined L^p-spaces, one sometimes encounters various generalizations. For instance, the Lebesgue measure can be replaced by another measure (see A. Friedman [1]). Also, one can use a weight function $g : [t_0, T] \to \mathbb{R}_+$, and define L^p_g by means of the condition

$$\int_{t_0}^T |x(t)|^p g(t)\, dt < +\infty.$$

We shall conclude this section by considering the space of *absolutely continuous* functions $x : [t_0, T] \to \mathbb{R}^n$. Convergence in this space is defined by means of convergence in $L^1([t_0, T], \mathbb{R}^n)$, and this is the reason we are dealing with this concept here.

From real analysis (see A. Friedman [1] or E.J. McShane [1]), it is known that an absolutely continuous function is a.e. differentiable, its derivative is an L^1-function, and Newton's formula $x(t) - x(t_0) = \int_{t_0}^t \dot{x}(s)\, ds$ is valid.

The set of absolutely continuous function is a linear space, complete with respect to the norm

$$|x(t)|_{AC} = |x(t_0)| + \int_{t_0}^T |\dot{x}(s)|\, ds.$$

The space of absolutely continuous functions from $[t_0, T]$ into $\mathbb{R}^n$, with the above norm, is a *Banach space*, and it is denoted by $AC([t_0, T], \mathbb{R}^n)$. It is almost obvious that

$$C^{(1)}([t_0, T], \mathbb{R}^n) \subset AC([t_0, T], \mathbb{R}^n).$$

2.3 Operators on function spaces

Most operators defined on function spaces have been defined and investigated in connection with their use in the theory of functional or functional differential equations. For instance, related to the study of ordinary differential equations of the form $\dot{x}(t) = f(t, x(t))$, the operator

$$(Nx)(t) = f(t, x(t)), \tag{43}$$

usually called the *Niemytskii operator*, has been thoroughly investigated by many authors (for some references, see C. Corduneanu [10]).

If one assumes $f(t, x)$ to be continuous for $(t, x) \in [t_0, T] \times \mathbb{R}^n$, then (43) defines an operator on the space $C([t_0, T], \mathbb{R}^n)$. This means that for each $x(t) \in C([t_0, T], \mathbb{R}^n)$, one has $(Nx)(t) \in C([t_0, T], \mathbb{R}^n)$. It is an elementary exercise to show that N is a *continuous* operator on C.

The operator N, defined by (43), has been also considered by C. Carathéodory, who (without any concern for operator theory) has proven a result which plays a basic role when looking for absolutely continuous solutions of $\dot{x}(t) = f(t, x(t))$. Of course, for such solutions the equation will be satisfied only almost everywhere (an absolutely continuous function is almost everywhere differentiable).

Carathéodory's result, related to the operator N defined by (43), can be stated as follows: Let $f : [t_0, T] \times \mathbb{R}^n \to \mathbb{R}^n$ be measurable in t, for fixed x, and continuous in x for almost all t. Then the operator N acts on the space of measurable (Lebesgue) functions from $[t_0, T]$ into $\mathbb{R}^n$, i.e., takes any measurable function into another measurable function.

The proof of Carathéodory's result can be found in E.J. McShane [1]. The main goal of this section is to define and provide some basic properties of *abstract Volterra operators*, or, as they are usually called in the engineering literature, *causal operators*. The term *nonanticipative operators* is also used.

In the Introduction (Chapter 1), we briefly discussed the concept of a causal operator, and illustrated this concept with examples from the classical literature.

Let $E = E([t_0, T), \mathbb{R}^n)$ and $F = F([t_0, T), \mathbb{R}^n)$ be two function spaces (usually consisting of continuous or measurable functions). Denote by $V : E \to E$ a map satisfying the condition (of causality!):

If $x, y \in E$ are such that $x(s) = y(s)$ on $t_0 \leq s \leq t < T$, then $(Vx)(s) = (Vy)(s)$ on the same interval $[t_0, T)$, with arbitrary $t < T$.

This definition is adequate when E and F are spaces of continuous functions. In case we deal with spaces of measurable functions, we must require the equalities above to be satisfied only almost everywhere.

The above definition shows clearly that the values $(Vx)(s)$ takes, for $s \leq t$, are determined by the values of $x(s)$ for $s \leq t$. Hence, the term *nonanticipative* is also justified. The examples provided in the Introduction constitute concrete cases of causal operators appearing in various functional equations intervening in the description of real phenomena.

The following properties of causal operators are elementary, and we are listing them here in view of their use in subsequent sections. We restrict our consideration to operators acting on a given space $E([t_0, T), \mathbb{R}^n)$, i.e., defined on this space and taking their values in the space: $V: E \to E$.

(1) The sum and the product of two causal operators $V : E \to E$ and $W : E \to E$ are also causal. In other words, $((V + W)x)(t) = (Vx)(t) + (Wx)(t)$, and $((V \circ W)x)(t) = (V(Wx))(t)$ are causal.

This property of causal operators on a given function space $E([t_0, T), \mathbb{R}^n)$ shows that the set of causal operators on E can be organized as an algebra with respect to the operations $+$ (addition) and $\circ$ (composition), over the field of reals. More generally, instead of the field of reals one could use a function (scalar) field, due to the fact that $a(t)(Vx)(t) = (V_1x)(t)$ is a causal operator if $a(t)$ is a real-valued function.

In particular, one can consider only linear causal operators on E, and organize them as a subalgebra of the algebra of all causal operators on E.

(2) If $V : E \to E$ is a causal operator, then the following operators on E are also causal:

$$(V_1x)(t) = a(t)(Vx)(t) + b(t),$$

$$(V_2x)(t) = a(t)|(Vx)(t)|(Vx)(t) + b(t)(Vx)(t) + c(t),$$

where $a(t)$, $b(t)$ and $c(t)$ are scalar functions. In an obvious manner one can consider "higher degree polynomials" in V.

(3) Sometimes, it is relatively easy to establish some connections between causal (abstract Volterra) operators and the classical Volterra operators.

For instance, if $K : L^2([t_0, T], \mathbb{R}) \to L^2([t_0, T], \mathbb{R})$ is a linear causal operator, then

$$(Lx)(t) = \int_{t_0}^{t} (Kx)(s)\, ds, \quad t \in [t_0, T], \tag{44}$$

is obviously linear and causal on the same space. As pointed out by C. Corduneanu [10], Chapter 2, L can be represented in the form

$$(Lx)(t) = \int_{t_0}^{t} k(t, s)x(s)\, ds, \quad t \in [t_0, T],$$

for some convenient kernel k. The case of $L^2([t_0, T], \mathbb{R}^n)$ is similar.

(4) A remarkable result establishing the connection between causal/ abstract Volterra operators and classical (integral) Volterra operators was obtained by I.W. Sandberg [1], [2], [3]. It states that a continuous causal operator on the space of scalar measurable functions can be represented by means of a Volterra series:

$$(Vx)(t) = f(t) + \sum_{m=1}^{\infty} \underbrace{\int_{t_0}^{t} \cdots \int_{t_0}^{t}}_{m \text{ times}} k_m(t_1, \ldots, t_m)x(t - t_1) \cdots x(t - t_m)\, dt_1, \ldots, dt_m.$$

We will not discuss here the details of the validity of the above representation, like the convergence of the series, properties of the kernels k_m etc., referring the reader to the above-mentioned references.

It is worthwhile pointing out that the so–called Volterra series play a significant role in system engineering theories. For details on this matter we refer the reader to the books by W.J. Rugh [1] and M. Schetzen [1]. These books deal with the Volterra–Wiener approach in system theory.

(5) Another property of causal operators is related to the convergence of a sequence of such operators. For illustration, let us take the space $C([t_0, T), \mathbb{R}^n)$ as the underlying space. Let $\{V_m; \ m \geq 1\}$ be a sequence of causal operators on C, such that

$$\lim(V_m x)(t) = (Vx)(t) \quad \text{as } m \to \infty, \tag{45}$$

for each fixed (t, x), $t \in [t_0, T]$, $x \in C$. Let us point out that convergence in (45) is weaker than convergence in C.

The problem is whether we can infer that $V : C \to C$ is also a causal operator. In other words, if $x, y \in C$ and $x(s) = y(s)$ on $t_0 \leq s \leq t$, does this property imply $(Vx)(s) = (Vy)(s)$ on $[t_0, t]$?

The answer is affirmative because the causality of V_m, $m \geq 1$, implies

$$(V_m x)(s) = (V_m y)(s), \quad s \in [t_0, T]. \tag{46}$$

If we let $m \to \infty$ in both sides of (46), and rely on (45), for fixed s, one obtains the causality of V.

(6) We have to consider now the problem of invertibility of causal operators, due to its significance in the study of functional equations involving such operators.

Generally speaking, the invertibility of causal operators is not possible within this class. The following is a rather simple example. We choose as underlying space the space $C([0, \infty), \mathbb{R}^n)$, and on C the operator

$$(Vx)(t) = x(t/2), \quad t \in [0, \infty). \tag{47}$$

This operator is obviously acting on C, and it is continuous. It has an inverse in C, namely

$$(V^{-1}y)(t) = y(2t), \quad t \in [0, \infty). \tag{48}$$

Moreover, this operator is continuous on C (i.e., if $y_m \to y$ uniformly on any interval $[0, T]$, then $V^{-1}y_m \to V^{-1}y$ on any bounded interval of $\mathbb{R}_+$).

But, as seen from (48), V^{-1} is not a causal operator.

Sometimes, it is not possible to invert a causal operator on a given function space, but the inverse into a larger space does exist and it may or may not be a causal operator.

The above discussion shows that, in general, we cannot take for granted the causality of the inverse of a causal operator. This statement agrees with the well-known fact that in an algebra, only some of its elements are invertible.

Hence, the causality of the inverse of a causal operator must be postulated, even though the existence is assured.

On the other hand, with reference to the classical examples of causal operators, we can indicate many circumstances in which the inverse does exist and is also causal.

An example which has an important role in the investigation of the so-called linear integral equation (of Volterra) of the second kind

$$x(t) = \int_{t_0}^{t} k(t, s)x(s) \, ds + f(t), \quad t \in [t_0, T] \tag{49}$$

is provided by the operator

$$(Lx)(t) = x(t) - \int_{t_0}^{t} k(t, s)x(s) \, ds. \tag{50}$$

We shall choose the space $C([t_0, T], \mathbb{R}^n)$ as the underlying space, and assume that the kernel $k(t, s)$ is a continuous matrix (n by n) valued function on $t_0 \leq s \leq t \leq T$. The equation (49) can now be written as $(Lx)(t) = f(t), t \in [t_0, T]$. Inverting the operator L on the space C means solving (49), and expressing $x = L^{-1}f$ in terms of $k(t, s)$ and $f(t)$.

This can be achieved by means of the resolvent formula

$$x(t) = f(t) + \int_{t_0}^{t} R(t, s)f(s) \, ds, \tag{51}$$

where $R(t, s)$ is the *resolvent kernel* associated with $k(t, s)$. In order to construct $R(t, s)$, we shall proceed solving (49) by successive approximations. Namely, we start the process with the function $f(t) \in C$, and consider the sequence

$$x^{(1)}(t) = \int_{t_0}^{t} k(t, s) f(s) \, ds + f(t),$$

$$x^{(2)}(t) = \int_{t_0}^{t} k(t, s) x^{(1)}(s) \, ds + f(t),$$

$$\vdots$$

$$x^{(m)}(t) = \int_{t_0}^{t} k(t, s) x^{(m-1)}(s) \, ds + f(t),$$

$$\vdots$$

Let us notice that $x^{(m+1)}(t)$ can be also written as

$$x^{(m+1)}(t) = \int_{t_0}^{t} \left[\sum_{j=1}^{m} k_j(t, s) \right] f(s) ds + f(t), \tag{52}$$

where $k_1(t, s) = k(t, s)$ and

$$k_j(t, s) = \int_{s}^{t} k(t, u) k_{j-1}(u, s) \, du, \quad j \geq 2. \tag{53}$$

If we assume $K > 0$ is such that $|k(t, s)| \leq K$ in $t_0 \leq s \leq t \leq T$, the norm for k being the Euclidean norm, for instance, then (53) easily leads to the estimate

$$|k_j(t, s)| \leq K^j \frac{(t - s)^{j-1}}{(j - 1)!}, \quad j \geq 2. \tag{54}$$

These estimates show that the series

$$\sum_{j=1}^{\infty} k_j(t, s) = R(t, s) \tag{55}$$

is uniformly convergent in the triangle $t_0 \leq s \leq t \leq T$, and its sum $R(t, s)$ is continuous there. $R(t, s)$ is known as the *resolvent* kernel associated with $k(t, s)$.

The resolvent formula (51) is now easily obtained from (52), by letting $m \to \infty$. Of course, this formula provides a solution in $C([t_0, T], \mathbb{R}^n)$ of the linear Volterra equation (of the second kind) (49). That this solution is unique one can easily see from the following argument. If $x(t)$ and $y(t)$ are two solutions of (49), then

$$x(t) - y(t) = \int_{t_0}^{t} k(t, s) [x(s) - y(s)] \, ds. \tag{56}$$

By recurrence, (56) leads to

$$x(t) - y(t) = \int_{t_0}^{t} k_m(t, s) [x(s) - y(s)] \, ds, \quad m \geq 2. \tag{57}$$

If one takes into account the estimates (54), the conclusion is $x(t) - y(t) \equiv 0$ on $[t_0, T]$.

Returning to the operator L, given by (50), we can now state that

$$(L^{-1}f)(t) = f(t) + \int_{t_0}^t R(t,s)f(s)\,ds, \tag{58}$$

which proves the invertibility of L on the space $C([t_0, T], \mathbb{R}^n)$ and its causality.

It is obvious from formula (58) that L^{-1} is a *causal operator*. The invertibility problem can also be formulated in different spaces than $C([t_0, T], \mathbb{R}^n)$. A rather common case is when $k(t, s)$ is assumed only *measurable* in $t_0 \le s \le t \le T$, and satisfies

$$\int_{t_0}^T dt \int_{t_0}^t |k(t,s)|^2\,ds < +\infty. \tag{59}$$

Then, if $f \in L^2([t_0, T], \mathbb{R}^n)$, one can discuss the problem of invertibility in the space $L^2([t_0, T], \mathbb{R}^n)$. This problem also has an affirmative answer and for the details of the proof we refer the reader to F. Tricomi's book [1]. The resolvent kernel associated with $k(t, s)$ is (formally) constructed as above, the estimates for convergence being somewhat different.

Actually, the L^2-case, as well as more general cases, will be discussed in Chapter 4 when dealing with linear causal equations.

The true invertibility problem for linear integral operators is related to the study of the so-called Volterra equation of the first kind:

$$\int_{t_0}^t k(t,s)x(s)\,ds = f(t). \tag{60}$$

For a brief (elementary) discussion of (60), we refer the reader to our book C. Corduneanu [23]. Under suitable conditions, the inverse is also causal.

(7) Before concluding this section on operators on function spaces, and particularly on *causal* operators, we would like to emphasize the fact that these operators allow some sort of "localization".

If $V : E \to E, E = E([t_0, T), \mathbb{R}^n)$, is causal, then it is possible to localize V to the restrictions of the functions in E to some subinterval $[t_0, t_1] \subset [t_0, T)$. In other words, if E_1 consists of the restrictions of all functions in E to $[t_0, t_1]$, then V can be "localized" to the space $E_1([t_0, t_1], \mathbb{R}^n)$. The causality is obviously preserved and we are entitled to consider this operator as a "*localized*" version of V.

Moreover, assume a causal operator V is given on the function space $E([t_0, T), \mathbb{R}^n)$. In studying problems for functional equations involving causal operators (see Chapter 3), it will be necessary to deal with another type of localization of the operators involved. Before defining the new scheme, it is useful to have a look at the following (classical case). Consider the Volterra integral equation (1.8)

$$x(t) = f(t) + \int_{t_0}^t K(t,s,x(s))\,ds, \quad t \in [t_0, T),$$

and assume we have a solution $\bar{x}(t)$ defined on an interval $[t_0, t_1], t_1 < T$. If we want to continue this solution to a larger interval, say $[t_0, t_2], t_2 > t_1$, then it is natural to replace (1.8) by the equation

$$x(t) = f(t) + \int_{t_0}^{t_1} K(t,s,\bar{x}(s))\,ds + \int_{t_1}^t K(t,s,x(s))\,ds,$$

and try to find a (local) solution defined on some interval $[t_1, t_2]$, $t_2 > t_1$. It is obvious that such a solution, say $\overline{\overline{x}}(t)$, if it exists, together with $\bar{x}(t)$, leads to a solution $x(t)$ of (1.8), defined on $[t_0, t_2]$. Namely, $x(t) = \bar{x}(t)$ on $[t_0, t_1]$ and $x(t) = \overline{\overline{x}}(t)$ on $(t_1, t_2]$. When the underlying space $E([t_0, T), \mathbb{R}^n)$ consists of measurable functions, the fact that $\bar{x}(t_1) \neq \overline{\overline{x}}(t_1)$ does not cause any incovenience.

The above example suggests the following scheme of "localization" of a causal operator. If instead of (1.8) we deal with the general functional equation with causal operator (1.16), $x(t) = (Vx)(t)$, then having a local solution on some interval $[t_0, t_1]$, say $\bar{x}(t)$, one can redefine the operator V on the space $E([t_1, t_2], \mathbb{R}^n)$ in the following manner: for each $y \in E([t_0, t_2], \mathbb{R}^n)$ we let $y(t) = \bar{x}(t)$ on $[t_0, t_1]$, and leave it unchanged on $(t_1, t_2]$. While $(Vy)(t)$ is defined on $[t_0, t_2]$, it is obvious that $(Vy)(t) = (V\bar{x})(t)$ on $[t_0, t_1]$. Then we let $(\overline{V}x)(t) = (Vx)(t)$ on $[t_1, t_2]$, and consider the equation $x(t) = (\overline{V}x)(t)$ on $[t_1, t_2]$. Any solution of this equation, pieced together with $\bar{x}(t)$, provides a solution of (1.16) on $[t_0, t_2]$. $\overline{V}$ can be regarded as a "localization" of V.

2.4 Fixed points and other auxiliary results

One of the most useful tools in proving the existence of solutions of functional equations is related to the existence of fixed points to operators acting on function spaces. For instance, for the functional equation (1.16) $x(t) = (Vx)(t)$, it is obvious that a solution is a fixed point (element) for the operator V, i.e., the action of the operator leaves that point (element) invariant.

The most often used fixed point theorem is, likely, the Banach contraction principle. It can be stated as follows:

Banach contraction mapping principle   *Let (S, d) be a complete metric space, and $A : S \to S$ a contraction mapping:*

$$d(Ax, Ay) \leq \rho d(x, y), \tag{61}$$

where $0 < \rho < 1$, for each $x, y \in S$. Then, there exists a unique fixed point x of A in S: $Ax = x$.

Proof   Let $x_0 \in S$ be an arbitrary point. One forms the sequence (successive approximations or iterates) $\{x_k\}$, $x_{k+1} = Ax_k$, $k \geq 0$. It is obvious that $x_k = A^k x_0$, $k \geq 1$, which implies on behalf of (61) $d(x_k, x_{k+1}) \leq \rho^k d(x_0, Ax_0)$, $k \geq 1$. Hence, for $m < n$,

$$d(x_m, x_n) \leq d(x_m, x_{m+1}) + d(x_{m+1}, x_{m+2}) + \cdots + d(x_{n-1}, x_n)$$
$$\leq (\rho^m + \rho^{m+1} + \cdots + \rho^{n-1}) d(x_0, Ax_0)$$
$$\leq \frac{\rho^m}{1 - \rho} d(x_0, Ax_0).$$

Therefore, $\{x_k\}$ is a Cauchy sequence in S, which means it is convergent:

$$\lim_{k \to \infty} x_k = x. \tag{62}$$

Since (61) implies the continuity of A, one obtains from $x_{k+1} = Ax_k$, $x = Ax$. Also, (61) implies the uniqueness of x. Indeed, from $x = Ax$, $y = Ay$, one has $d(x, y) \leq \rho d(x, y)$, which is possible only if $d(x, y) = 0$.

From the above inequality for $d(x_m, x_n)$ one obtains (letting $n \to \infty$)

$$d(x_m, x) \leq \frac{\rho^m}{1 - \rho} d(x_0, Ax_0),\tag{63}$$

which provides an estimate for the "error" occuring when x_m is substituted for x.

Remark   The Banach contraction mapping principle provides an abstract setting for the classical method of iteration/successive approximations. Its success depends not only on the choice of the space S, but also on the choice of the metric function d.

Another fixed point result which has many applications in the theory of functional equations is Schauder's fixed point theorem. In order to state it, we need to define the concept of a *convex* set in a linear space.

Let X be a linear space and $C \subset X$ a set. We say that C is *convex* if the following property is valid: for any $x, y \in C$ and $m \in (0, 1)$, one has $mx + (1 - m)y \in C$.

Schauder's fixed point theorem   *Let B be a Banach space and $C \subset B$ a convex, closed and bounded set. If $T : B \to B$ is a continuous operator such that*

$$TC \subset C, \quad TC \text{ is relatively compact},\tag{64}$$

then T has at least one fixed point in C.

The proof of Schauder's fixed point theorem can be found in many books: see, for instance, C. Corduneanu [3], K. Deimling [1], L.V. Kantorovich and G.P. Akilov [1].

A more general result than Schauder's fixed point theorem is Tychonoff's fixed point theorem in Fréchet spaces. We are intentionally limiting ourselves to a special case.

Tychonoff's fixed point theorem   *Let F be a Fréchet space whose distance function is constructed by means of a sufficient countable family of seminorms. If $C \subset F$ is a closed convex set, and $T : C \to C$ is a continuous operator such that TC is relatively compact, then T has at least one fixed point in C.*

The proof of Tychonoff's theorem can be found, for instance, in C. Corduneanu [3], [10], R. Edwards [1]. The Fréchet space is actually replaced by a locally convex topological vector space, a concept we did not discuss in this book.

Let us point out the fact that Tychonoff's theorem will be used in getting global existence results.

Other fixed point theorems, or related tools of functional analysis necessary in proving existence theorems for functional equations, can be found in C. Corduneanu [3], [10], K. Deimling [1], R.E. Edwards [1].

We shall now discuss the concept of an *approximate unit* (or identity) for the convolution product. For a more detailed discussion, see C. Sadosky [1], for instance.

The convolution product of two functions $f \in L^1(R, R)$ and $g \in L^p(R, R)$, $1 \le p \le \infty$, is defined by

$$(f * g)(t) = \int_R f(t - s)g(s)\, ds. \tag{65}$$

$f * g$ is defined almost everywhere on $\mathbb{R}$, it belongs to L^p, and the following estimate is valid:

$$\|f * g\|_p \le \|f\|_1 \|g\|_p. \tag{66}$$

An *approximate unit* for the convolution product can be defined, for instance, by

$$\Phi_\varepsilon(t) = \begin{cases} 0, & t < 0, \\ \varepsilon^{-1}\exp(-\varepsilon^{-1}t), & t \ge 0, \end{cases} \tag{67}$$

where $\varepsilon > 0$ is arbitrary (usually, a small number). The term "approximate unit" is justified by the result

$$\lim(\Phi_\varepsilon * g) = g, \quad \text{as } \varepsilon \to 0, \tag{68}$$

for any $g \in L^p$, $1 \le p < \infty$. Convergence in (68) is meant in L^p.

If we notice that $\int_R \Phi_\varepsilon(t) = 1$ for $\varepsilon > 0$, then the proof of the above statement is a consequence of the formula

$$(f * \Phi_\varepsilon) - f(t) = \int_R [f(t - \varepsilon s) - f(t)]\Phi_1(s)\, ds,$$

to which one applies the inequality (66), after splitting the integral conveniently.

The case $p = \infty$ is not covered by the above statement. Related to this case, one can replace L^∞ by the space of uniformly continuous and bounded functions (which is a subspace of L^∞). Then (68) remains valid, with the convergence being uniform on $\mathbb{R}$.

3 Existence theory for functional equations with causal operators

3.1 The equation $x(t) = (Vx)(t)$ in the space of continuous functions

In Chapter 1 we observed that several types of functional equations encountered in classical mathematics and its applications, such as ordinary differential equations (1.1), equations with delayed argument (1.4) or (1.5), integral equations like (1.8), or integrodifferential equations like (1.10), can be regarded as special cases of the functional equation (1.16), i.e.,

$$x(t) = (Vx)(t), \tag{1}$$

in which V is a causal operator acting on an appropriate function space.

The case of a functional differential equation like (1.13)

$$\dot{x}(t) = (Vx)(t), \tag{2}$$

together with the usual initial condition, has been reduced (by integrating both sides of (2)) to the form (1), namely

$$x(t) = x^0 + \int_{t_0}^{t} (Vx)(s)\, ds. \tag{3}$$

Of course, this procedure can be applied any time V takes values in a space of measurable functions (locally integrable).

Therefore, for the existence of solutions to either (1) or (2), we can deal with equation (1) only.

Historically, the first existence theorem for equation (1) was established by L. Tonelli [1]. It is concerned with the space of continuous functions, as the underlying space for equation (1).

Theorem 3.1 *Let us consider the functional equation* (1), *in which V stands for a causal operator on the space $C([t_0, T], \mathbb{R}^n)$, continuous and compact. Moreover, assume V has the property of fixed initial value. Then, there exists a solution of* (1) *in $C([t_0, t_0 + \delta], \mathbb{R}^n)$, for some $\delta > 0$, where $t_0 + \delta \leq T$.*

Proof If $c \in \mathbb{R}^n$ denotes the fixed initial value of V, i.e., $(Vx)(t_0) = c$ for any $x \in C([t_0, T], \mathbb{R}^n)$, then we consider the equation

$$x(t) - c = (Vx)(t) - c, \tag{4}$$

which is (obviously) equivalent to (1). Due to the compactness of V, if we fix an $r > 0$, we can write $|(Vx)(t) - c| < \varepsilon$ for $t - t_0 < \delta$ and all $x \in C$ with $|x(t) - c| \leq r$, $t \in [t_0, T]$.

Once r is fixed, δ depends on ε only. Without loss of generality, we can assume $\varepsilon \leq r$. Then, (4) implies $|(Vx)(t) - c| < \varepsilon \leq r$ for all $x(t)$ with $|x(t) - c| \leq r$, as soon as $t \leq t_0 + \delta$.

In other words, the ball of radius r, centered at c in the space $C([t_0, t_0 + \delta], \mathbb{R}^n)$, is taken into itself by the operator V. According to our assumption, V is continuous and compact. Therefore, Schauder's fixed point theorem (see Chapter 2) applies in the Banach space $C([t_0, t_0 + \delta], \mathbb{R}^n)$. The ball $|x(t) - c| \leq r$ is obviously convex and closed.

Therefore, there exists at least one solution of (1), defined on the interval $[t_0, t_0 + \delta]$, and belonging to the space C.

This ends the proof of Theorem 3.1.

Remark   While the result of Theorem 3.1 is a *local* result (i.e., the solution is defined only in a "small" interval $[t_0, t_0 + \delta]$), the argument in the proof can be used to obtain a *global* existence result (i.e., the solution is defined on the whole interval $[t_0, T]$). To reach this goal, we must add further restrictions on the operator V. A rather strong condition is the (total) boundedness of V on the space $C([t_0, T], \mathbb{R}^n)$: $|(Vx)(t)| \leq M < \infty$, for $t \in [t_0, T]$ and any $x \in C$.

Then, based on this property for V, one can easily find an "a priori" estimate for the solution of (1), if any. Indeed, (1) shows that any solution of this equation, if defined on $[t_0, T]$, also satisfies $|x(t)| \leq M$, $t \in [t_0, T]$. Hence, if we consider the ball of radius M, centered at the origin of the space $C([t_0, T], \mathbb{R}^n)$, this ball is taken into itself by V. As a result of our assumptions, V has a fixed point in this ball. This means equation (1) has a continuous solution on $[t_0, T]$.

Corollary 3.2   *Consider the functional differential equation* (2), *under the initial condition* $x(t_0) = x^0 \in \mathbb{R}^n$. *Assume V is a causal operator on* $C([t_0, T], \mathbb{R}^n)$, *continuous and taking bounded sets into bounded sets. Then, there exists a continuously differentiable solution of the problem, defined on some interval* $[t_0, t_0 + \delta]$, $(t_0 + \delta \leq T)$.

Proof   Since the initial value problem (2), $x(t_0) = x^0$, is equivalent to equation (3), we need only to prove that the operator

$$(Wx)(t) = x^0 + \int_{t_0}^{t} (Vx)(s) \, ds, \tag{5}$$

appearing in the right-hand side of (3), satisfies the conditions required for V in Theorem 3.1.

The causality and continuity of W are almost obvious. Since convergence in C is uniform convergence, the continuity of V implies the continuity of W.

The compactness of W on C can be established as follows. Assume $x \in B \subset C$, with V bounded in C. According to our assumption, B is taken by V into a bounded set in C. Hence, for each $x \in B$ one has $|(Vx)(t)| \leq K$, $t \in [t_0, T]$. Therefore, for each $x \in B$ one has

$$|(Vx)(t)| \leq |x^0| + K(T - t_0), \quad t \in [t_0, T]$$

and

$$|(Wx)(t) - (Wx)(u)| \leq K|t - u|, \quad t, u \in [t_0, T].$$

These inequalities prove that the subset of C consisting of those functions, which are of the form $(Wx)(t)$, $x \in B$, satisfies the conditions required in the Ascoli–Arzelà criterion of compactness in $C([t_0, T], \mathbb{R}^n)$.

This ends the proof of Corollary 3.2.

Let us now apply Theorem 3.1 and its Corollary 3.2 to obtain existence results for some classical functional equations.

First, consider the *classical Volterra integral equation* (1.8)

$$x(t) = f(t) + \int_{t_0}^{t} K(t, s, x(s)) \, ds, \quad t \in [t_0, T], \tag{6}$$

under the following assumptions:

1   $f(t) \in C([t_0, T], \mathbb{R}^n)$;
2   $K(t, s, x)$ is continuous from $\Delta \times \mathbb{R}^n$ into $\mathbb{R}^n$, where $\Delta = \{(t, s); \ t_0 \leq s \leq t \leq T\}$.

Then there exists $\delta > 0$, $t_0 + \delta \leq T$, such that (6) admits a solution in $C([t_0, t_0 + \delta], \mathbb{R}^n)$.

The proof is a consequence of the fact that the operator

$$(Vx)(t) = f(t) + \int_{t_0}^{t} K(t, s, x(s)) \, ds \tag{7}$$

satisfies on $C([t_0, T], \mathbb{R}^n)$ the conditions required by Theorem 3.1. The continuity and causality of V are obvious. Compactness requires some elaboration but easily follows from elementary results of classical analysis.

Indeed, assume $x(t) \in B \subset C$, with B a bounded set in C. This means we can find $M > 0$, with the property $|x(t)| \leq M$, $t \in [t_0, T]$, $x \in B$. The set $\Delta \times \{x \in \mathbb{R}^n; \ |x| \leq M\}$ is compact (bounded and closed) in $\mathbb{R}^{n+2}$, which implies the boundedness of K on that set, $|K(t, s, x)| \leq A$ for some $A > 0$, as well as its uniform continuity.

From (7) one derives for $x \in B$

$$|(Vx)(t)| \leq \sup |f(t)| + A(T - t_0), \quad t \in [t_0, T],$$

and

$$|(Vx)(t + h) - (Vx)(t)| \leq |f(t + h) - f(t)|$$
$$+ \int_{t_0}^{t} [|K(t + h, s, x(s)) - K(t, s, x(s))|] \, ds$$
$$+ \int_{t}^{t+h} |K(t + h, s, x(s))| \, ds.$$

The first inequality above proves the uniform boundedness of the set $\{Vx; \ x \in B\}$, while the second immediately leads to the conclusion that this set is uniformly equicontinuous on $[t_0, T]$. The Ascoli–Arzelà criterion applies, which ends the proof of existence for (6).

A more general existence result can be obtained for equation (6), still looking for continuous solutions. See R.K. Miller [1] or C. Corduneanu [21].

Peano's theorem concerning the (local) existence of solutions to the problem $\dot{x}(t) = f(t, x(t))$, $x(t_0) = x^0 \in \mathbb{R}^n$, is obtained from the above existence result for (6), taking $f(t) = x^0$ and $K(t, s, x) = f(s, x)$.

Many other existence results can be obtained from Theorem 3.1 or its corollary.

We shall now consider the case of *delay equations* (1.5), namely

$$\dot{x}(t) = f(t, x_t), \tag{8}$$

where $x_t(s) = x(t+s)$, $-h \leq s \leq 0$, under the initial condition (1.7), i.e.,

$$x(s) = x_0(s), \quad s \in [t_0 - h, t_0] \tag{9}$$

or

$$x_{t_0} = x_0 \in C([t_0 - h, t_0], \mathbb{R}^n).$$

The following conditions will assure the existence of solutions:

1    $f(t, \varphi)$ is a continuous map from the product $[t_0, T] \times C([t_0 - h, t_0], \mathbb{R}^n)$ into $\mathbb{R}^n$;
2   f takes bounded sets (in its domain) into bounded sets in $\mathbb{R}^n$;
3    $x_0(s) \in C([t_0 - h, t_0], \mathbb{R}^n)$.

Then, the initial value problem (8), (9) has a solution in the space $C([t_0, t_0 + \delta], \mathbb{R}^n)$, for some $\delta > 0$, $t_0 + \delta \leq T$.

The proof can be carried out easily, first transforming our problem into an equivalent functional equation, namely

$$x(t) = x_0(t_0) + \int_{t_0}^{t} f(s, x_s) \, ds. \tag{10}$$

From (10) one sees that the operator

$$(Vx)(t) = x_0(t_0) + \int_{t_0}^{t} f(s, x_s) \, ds, \tag{11}$$

defined on the space $C([t_0, T], \mathbb{R}^n)$, is continuous and causal. To prove that it takes bounded sets into relatively compact sets, we will rely on the boundedness of f on any bounded set of $C([t_0, T], \mathbb{R}^n)$. It is the same kind of argument we have used in the proof of Corollary 3.2 above.

Hence, the initial value problem (8), (9) has a local solution (which is obviously continuously differentiable) on some interval $[t_0, t_0 + \delta]$.

The delay equations of the form (1.4), or even more general like

$$\dot{x}(t) = f(t, x(t - h_1), \ldots, x(t - h_m)), \tag{12}$$

where $0 \leq h_1 < h_2 < \cdots < h_m$, are particular cases of equation (8). The initial data must be assigned on the interval $[t_0 - h_m, t_0]$. The continuity assumption on f will suffice for the existence of a local solution.

Another case which is covered by Corollary 3.2 is provided by the integrodifferential equations

$$\dot{x}(t) = f\left(t, x(t), \int_{t_0}^{t} k(t, s, x(s)) \, ds\right). \tag{13}$$

One can associate with (13) the usual initial condition $x(t_0) = x^0 \in \mathbb{R}^n$. The following conditions imposed on f and k will assume the applicability of Corollary 3.2.

1   f is a continuous map from $[t_0, T] \times \mathbb{R}^n \times \mathbb{R}^n$ into $\mathbb{R}^n$;
2   k is a continuous map from $\Delta \times \mathbb{R}^n$ into $\mathbb{R}^n$.

We leave to the reader the task of checking the validity of the hypotheses in Corollary 3.2, on behalf of conditions 1 and 2 above. Let us recall that $\Delta = \{(t, s); \, t_0 \leq s \leq t \leq T\}$.

In the subsequent section we shall provide an existence theorem for (13), using measurability conditions.

In concluding this section, we will consider the functional equation

$$x(t) = f(t) + \int_{t_0}^{t} (Vx)(s)\,ds, \quad t \in [t_0, T], \tag{14}$$

which is slightly more general than equation (13), which in turn is equivalent to equation (2), under the initial condition $x(t_0) = x^0 \in \mathbb{R}^n$. Equation (14) will enable us to describe a method used by L. Tonelli in existence problems. This method is particularly adequate when an integration occurs in the equation.

In regard to equation (14) we will make the following hypotheses:

1   $f \in C([t_0, T], \mathbb{R}^n)$;
2   V is a causal operator on the space $C([t_0, T], \mathbb{R}^n)$, bounded on the ball

$$|x(t) - f(t)| \leq b, \tag{15}$$

i.e., there exists $M > 0$ such that

$$|(Vx)(t)| \leq M, \quad t \in [t_0, T], \tag{16}$$

for any $x(t)$ satisfying (15).

Then, there exists a solution of (14), defined on the interval $[t_0, t_0 + \delta]$, with

$$\delta = \min \left\{ T - t_0, \frac{b}{M} \right\}. \tag{17}$$

Tonelli's method is based on the construction of some approximate solutions to equation (14). Let us consider a partition $t_0 < t_1 < \cdots < t_p = t_0 + \delta$ of the interval $[t_0, t_0 + \delta]$, with $t_j = t_0 + jh$, $j = 1, 2, \ldots, p$, where $\delta = ph$. Then, consider the following sequence of continuous functions on $[t_0, t_0 + \delta]$, with values in $\mathbb{R}^n$: $x^{(1)}(t) = f(t)$, and for $p > 1$

$$x^{(p)}(t) = \begin{cases} f(t), & t \in [t_0, t_1], \\ f(t) + \displaystyle\int_{t_0}^{t-h} (Vx^{(p)})(s)\,ds, & t \in [t_1, t_0 + \delta]. \end{cases} \tag{18}$$

This is a step by step construction method. It should be understood as follows: having $x^{(p)}(t)$ on $[t_0, t_1]$, the formula (18) allows one to determine $x^{(p)}(t)$ on $[t_1, t_2]$. More precisely, on $[t_1, t_2]$, $x^{(p)}(t)$ is given by

$$x^{(p)}(t) = f(t) + \int_{t_0}^{t-h} (Vf)(s)\,ds.$$

Then one uses the values of $x^{(p)}(t)$ on $[t_1, t_2]$ to determine this function on $[t_2, t_3]$, and so on. In order to apply this procedure, we must be sure that

$$|x^{(p)}(t) - f(t)| \leq b, \quad t \in [t_0, t_0 + \delta]. \tag{19}$$

Since (19) is obviously valid on $[t_0, t_1]$, we will now assume it is valid on $[t_0, t_k]$, $k > 1$, and prove that it remains valid on $[t_k, t_{k+1}]$. Indeed, from (18) one obtains for $t \in [t_k, t_{k+1}]$

$$|x^{(p)}(t) - f(t)| \leq M(t_{k+1} - t_0 - h) \leq M\delta \leq b.$$

Hence, Tonelli's method leads to a sequence $\{x^{(p)}(t); \; p \geq 1\}$ of functions defined on $[t_0, t_0 + \delta]$. It is an easy exercise to prove that these functions are continuous at each t_k, $k = 1, 2, \ldots, p$, which means they are continuous on $[t_0, t_0 + \delta]$.

The sequence $\{x^{(p)}(t); \; p \geq 1\}$ is uniformly bounded on $[t_0, t_0 + \delta]$. In order to apply the Ascoli–Arzelà criterion of compactness, it remains to show that they are also equicontinuous on the same interval. The equicontinuity of the function $x^{(p)}(t)$, $p \geq 1$, follows from the inequality

$$|x^{(p)}(t) - x^{(p)}(s)| \leq |f(t) - f(s)| + M|t - s|,$$

which is a direct consequence of (18).

Using the Ascoli–Arzelà criterion, we can extract from $\{x^{(p)}(t); \; p \geq 1\}$ a subsequence, say $\{x^{(p_m)}(t); \; m \geq 1\}$, which converges uniformly on $[t_0, t_0 + \delta]$. Writing (18) for $p = p_m$, and letting $m \to \infty$, one obtains (14) where $x(t) = \lim x^{(p_m)}(t)$ as $m \to \infty$. This ends the proof of the existence for equation (14).

Remark   One can obtain results on the existence of solutions to the functional equation (14) without the (main) assumption that the operator V acts on the space $C([t_0, T], \mathbb{R}^n)$, with values in the same space. A more general assumption can also assure the existence of continuous solutions. To illustrate this assertion, we will make the following hypotheses on the data in equation (14):

1   $f(t) \in C([t_0, T], \mathbb{R}^n)$;
2   V is a continuous causal operator defined on $C([t_0, T], \mathbb{R}^n)$, with values in $L^1([t_0, T], \mathbb{R}^n)$;
3   there exists a ball in $C([t_0, T], \mathbb{R}^n)$,

$$|x(t) - f(t)| \leq b, \quad t \in [t_0, T], \tag{20}$$

and a function $h(t) \in C([t_0, T], \mathbb{R}_+)$, such that

$$|(Vx)(t)| \leq h(t), \quad \text{a.e. on } [t_0, T], \tag{21}$$

for each $x(t)$ satisfying (20).

Then, there exists a continuous solution of equation (14) on some interval $[t_0, t_0 + \delta]$, $t_0 + \delta \leq T$.

The proof of the above assertion can be carried out by means of Schauder's fixed point theorem in the space $C([t_0, t_0 + \delta], \mathbb{R}^n)$, where the number δ is the largest, $t_0 + \delta \leq T$, such that

$$\int_{t_0}^{t_0+\delta} h(s)\, ds \leq b. \tag{22}$$

Indeed, the ball

$$|x(t) - f(t)| \leq b, \quad t \in [t_0, t_0 + \delta] \tag{23}$$

belongs to the space $C([t_0, t_0 + \delta], \mathbb{R}^n)$, is closed, bounded and convex. The operator

$$(Wx)(t) = f(t) + \int_{t_0}^{t} (Vx)(s) \, ds \qquad (24)$$

takes the ball (23) into itself. From (24) one obtains

$$|(Wx)(t) - f(t)| \leq \int_{t_0}^{t_0+\delta} h(s) \, ds \leq b.$$

The causality of W is an immediate consequence of that of V, as well as its continuity:

$$|(Wx)(t) - (Wy)(t)| \leq \int_{t_0}^{t_0+\delta} |(Vx)(s) - (Vy)(s)| \, ds.$$

The compactness of W follows from the inequality

$$|(Wx)(t) - (Wx)(u)| \leq \left| \int_{u}^{t} h(s) \, ds \right|,$$

which shows the equicontinuity of the set $\{Wx\}$, with x in the ball (23). The Ascoli–Arzelà criterion leads to the conclusion.

In the special case $f(t) = x^0 \in \mathbb{R}^n$, $(Vx)(t) = g(t, x(t))$, where $g(t, x)$ is measurable in t for fixed x and continuous in x for almost all $t \in [t_0, T]$, equation (14) becomes

$$x(t) = x^0 + \int_{t_0}^{t} g(s, x(s)) \, ds,$$

which is equivalent to $\dot{x}(t) = g(t, x(t))$, a.e. on $[t_0, T]$, under the initial condition $x(t_0) = x^0$. In other words, for contant f and V an operator of Niemystkii type, satisfying Carathéodory's conditions, the existence result for (14) implies the classical Carathéodory theorem of existence for ordinary differential equations. As noticed above, these solutions are absolutely continuous functions (integrals of L^1-functions), satisfying the differential equation almost everywhere.

3.2 The equation $x(t) = (Vx)(t)$ in spaces of measurable functions

As usual, we will assume $x(t)$ takes its values in $\mathbb{R}^n$, while V is a causal operator acting on some space $E([t_0, T], \mathbb{R}^n)$, consisting of measurable functions.

Let us consider equation (1), choosing the space $L^p([t_0, T], \mathbb{R}^n)$, $1 \leq p < \infty$, as the underlying space. It is expected that continuity and compactness of the operator V have local existence as a consequence. Unlike the case of spaces of continuous functions, the "fixed initial value property" will not be required (actually, the elements of L^p being classes of measurable functions which differ on a zero measure set, this property does not make sense).

More precisely, the following existence result is valid for equation (1), in L^p-spaces, $1 \leq p < \infty$.

Theorem 3.3 *Consider equation (1), with V causal, continuous and compact on the space $L^p([t_0, T], \mathbb{R}^n)$, $1 \leq p < \infty$. Then, there exists $\delta > 0, t_0 + \delta \leq T$, such that (1) has a solution $x(t) \in L^p([t_0, t_0 + \delta], \mathbb{R}^n)$.*

Proof   The conditions imposed on V assure those required in Schauder's fixed point theorem. We need only to show that there exists a closed convex set in L^p, which is taken into itself by the operator V. While this may not be possible in $L^p([t_0, T], \mathbb{R}^n)$, it is always possible in $L^p([t_0, t_0 + \delta], \mathbb{R}^n)$, with sufficiently small δ. Actually, the closed convex set can be chosen to be a ball, centered at the origin of the space L^p.

Indeed, let us fix a number $r > 0$ and consider

$$B_r = \{x; \; x \in L^p, \; |x_p| \le r\}.$$

We need to show that for sufficiently small δ, one has $V B_r \subset B_r$. This means

$$\int_{t_0}^{t_0+\delta} |(Vx)(t)|^p \, dt \le r^p, \quad \text{for each } x \in B_r. \tag{25}$$

First, let us notice that the relative compactness of $V B_r$ implies the existence of a finite ε-net, for every $\varepsilon > 0$. Given $\varepsilon > 0$, let $v_1, v_2, \ldots, v_m$ denote the elements of $V B_r$, such that $\{V v_j\}$, $j = 1, 2, \ldots, m$, is the finite ε-net of $V B_r$. For $x = v_1, v_2, \ldots, v_n \in B_r$, we can write

$$\int_{t_0}^{t_0+\delta} |(V v_j)(t)|^p \, dt < \varepsilon^p, \quad j = 1, 2, \ldots, m, \tag{26}$$

provided we choose δ small enough. When $\delta \to 0$, the left-hand side of (26) tends to zero for each j, $j = 1, 2, \ldots, m$, and since we have only finitely many v_j, the assertion is obviously true. If one now takes an arbitrary $u \in V B_r$, then we can find a j, $j = 1, 2, \ldots, m$, such that $|u - V v_j|_p < \varepsilon$, according to the definition of an ε-net (see Chapter 2). Since for some j, $u = (u - V v_j) + V v_j$, we obtain by means of (26)

$$\int_{t_0}^{t_0+\delta} |u(t)|^p \, dt \le (\varepsilon + \varepsilon)^p = 2^p \varepsilon^p, \tag{27}$$

which compared to (25) leads to the conclusion that we must take $\varepsilon < (r/2)$. Of course, this is not a restriction, given the meaning of ε.

Therefore, if one chooses $\delta = \delta(\varepsilon)$ such that (26) takes place, then (25) holds true and this ends the proof of Theorem 3.3.

Remark   The case $p = \infty$ is not dealt with in Theorem 3.3. In Theorem 3.1 we treated, under an extra condition on the operator V (namely, the "fixed initial value property"), the case of the space $C([t_0, T], \mathbb{R}^n)$. This is a subspace of $L^\infty([t_0, T], \mathbb{R}^n)$. In a subsequent chapter, we shall also consider the case of L^∞ as the underlying space.

From Theorem 3.3 we can derive an existence result for the functional differential equation (2), under the usual initial condition $x(t_0) = x^0$.

Corollary 3.4   *Let us consider the functional differential equation (2), under the initial condition $x(t_0) = x^0$, and assume V satisfies the following conditions:*

1   *V is a causal continuous operator on the space $L^p([t_0, T], \mathbb{R}^n)$, $1 \le p < \infty$;*
2   *V is bounded, i.e., takes bounded sets in L^p into bounded sets.*

Then, there exists a solution of equation (2), satisfying $x(t_0) = x^0$, and such that $x \in AC([t_0, t_0 + \delta], \mathbb{R}^n)$, for some $\delta > 0$, $(t_0 + \delta \leq T)$. Moreover, $\dot{x}(t) \in L^p([t_0, t_0 + \delta], \mathbb{R}^n)$.

Proof As seen above, equation (2) with initial condition $x(t_0) = x^0$ leads to the functional equation $x(t) = (Wx)(t)$, where W is given by (5). Therefore, it suffices to check the hypotheses of Theorem 3.3 for the operator W. Causality and continuity are evident. It remains to prove that W is compact on L^p. But the set $\{Wx; \ x \in B\}$, where B is a bounded set in L^p, consists of absolutely continuous functions (hence, continuous). It is easy to prove that $\{Wx; \ x \in B\}$ is a compact set in the space $C([t_0, T], \mathbb{R}^n)$. Indeed, in case $p > 1$ we have for $y = Wx$, $t, u \in [t_0, T]$

$$|y(t) - y(u)| \leq \left| \int_u^t |(Vx)(s)| \, ds \right|,$$

which implies

$$|y(t) - y(u)| \leq |t - u|^{1/q} \left(\int_u^t |(Vx)(s)|^p ds \right)^{1/p}.$$

This inequality shows the equicontinuity of the functions in $\{Wx; \ x \in B\}$ on $[t_0, T]$. Since convergence in C implies convergence in L^p, there follows the compactness of $\{Wx; \ x \in B\}$ in the space L^p.

The case $p = 1$ is not covered by the above argument. It will be dealt with in the next section, when we present the singular perturbation approach.

We shall now consider some applications of Theorem 3.3, or its Corollary 3.4, to some classical types of function equations.

First, we shall consider the nonlinear integral equation of Hammerstein–Volterra type:

$$x(t) = f(t) + \int_{t_0}^t k(t, s)g(s, x(s)) \, ds, \quad t \in [t_0, T], \tag{28}$$

in which x and f take values in $\mathbb{R}^n$, $k(t, s)$ is a matrix kernel of type $n \times n$, while g is a map from $[t_0, T] \times \mathbb{R}^n$ into $\mathbb{R}^n$. The following conditions on f, k and g will assure the local existence of the L^p-solution:

1 $f \in L^p([t_0, T], \mathbb{R}^n)$, $1 < p < \infty$;
2 k is measurable on $\Delta = \{(t, s); \ t_0 \leq s \leq t \leq T\}$, and such that, with $p^{-1} + q^{-1} = 1$,

$$\left\{ \int_{t_0}^T \left(\int_{t_0}^t |k(t, s)|^p \, ds \right)^{q/p} dt \right\}^{1/q} = M < \infty, \tag{29}$$

where $|k(t, s)|$ is any matrix norm (for instance, the Euclidean norm $|k| = (\sum_{i,j=1}^n |k_{ij}|^2)^{1/2}$);
3 g is measurable on $[t_0, T] \times \mathbb{R}^n$, and such that $x(t) \to g(t, x(t))$ maps $L^p([t_0, T], \mathbb{R}^n)$ into itself continuously, taking bounded sets into bounded sets.

Then, there exists a solution $x \in L^p([t_0, t_0 + \delta], \mathbb{R}^n)$ of equation (28), for some $\delta > 0$ $(t_0 + \delta \leq T)$.

The proof follows from Theorem 3.3, if we show that the operator

$$(Ux)(t) = f(t) + \int_{t_0}^{t} k(t,s)g(s,x(s))\,ds, \tag{30}$$

acts continuously on $L^p([t_0, t_0 + \delta], \mathbb{R}^n)$, and is a compact operator. U is obviously causal. Both continuity and compactness are consequence of assumptions 2 and 3. It suffices to notice that condition 2 assures the continuity and compactness of the linear integral operator

$$x(t) \to \int_{t_0}^{t} k(t,s)x(s)\,ds \tag{31}$$

on the space $L^p([t_0, t_0 + \delta], \mathbb{R}^n)$. See, for instance, C. Corduneanu [3].

One has also to take into account that the product (superposition) of the operators $x(t) \to g(t, x(t))$ and (31) is continuous and compact on $L^p([t_0, t_0 + \delta], \mathbb{R}^n)$.

Remark   Condition 3 can be satisfied in various ways. For intance, one can assume $|g(t, x)| \leq M|x| + g_0(t)$, $M > 0$, and g_0 nonnegative in L^p.

Remark   It is interesting to point out that condition (29), for $p = 2$, becomes the square integrability of k (which is known, for linear Fredholm integral operators, as the Hilbert–Schmidt condition).

The more general integral equation (6) can also be dealt with in L^p-space. But the continuity and compactness conditions in the general nonlinear case are more difficult to handle. In the particular case $p = 2$, i.e., of square integrable solutions, this equation is elegantly treated in F. Tricomi's book [1]. The existence (global) and uniqueness are proved by successive approximations. See also G. Gripenberg *et al.* [1] for the general case (any $p \geq 1$), and equation (1).

A second application of the general results in this section is concerned with the *integrodifferential* equation

$$\dot{x}(t) = f\left(t, x(t), \int_{t_0}^{t} k(t,s)x(s)\,ds\right), \tag{32}$$

under the usual initial condition $x(t_0) = x^0 \in \mathbb{R}^n$. Such equations are often encountered in applications, sometimes as perturbed equations of ordinary differential equations, when it has the (special) form

$$\dot{x}(t) = f(t, x(t)) + \int_{t_0}^{t} k(t,s)x(s)\,ds.$$

As noticed in Chapter 1, equation (1.11), this kind of perturbation is usually related to the "memory" of the system.

Consider now equation (32), which is a particular case of equation (1.10), in the sense that the integral term is linear. This simplifies somewhat the statement of conditions (due to the better knowledge we have about linear integral operators, as oppposed to nonlinear ones).

We will assume that the following hypotheses are satisfied by the function $f(t, x, y)$, occuring in equation (32).

1   $f : [t_0, T] \times \mathbb{R}^n \times \mathbb{R}^n \to \mathbb{R}^n$ is of Carathéodory type, i.e., measurable in t for each fixed x and y, and continuous with respect to x and y for almost all $t \in [t_0, T]$;

2   there exist a constant $M > 0$ and a nonnegative function $g \in L^2([t_0, T], \mathbb{R})$, such that

$$|f(t, x, y)| \leq M(|x| + |y|) + g(t), \tag{33}$$

   for $t \in [t_0, T]$ and $x, y \in \mathbb{R}^n$;
3    $k(t, s) = (k_{ij}(t, s))_{n \times n}$ is measurable on $\Delta = \{(t, s); t_0 \leq s \leq t \leq T\}$, and satisfies condition (29).

   Then, the integrodifferential equation (32) has a solution $x(t)$, satisfying a.e. the equation and the initial condition $x(t_0) = x^0$. This solution is absolutely continuous on the existence interval $[t_0, t_0 + \delta]$, $\delta > 0$, $(t_0 + \delta \leq T)$, while $\dot{x}(t) \in L^2([t_0, t_0 + \delta], \mathbb{R}^n)$.
   The proof of the above assertion follows from Corollary 3.4. Indeed, we have to prove that the operator

$$(Wx)(t) = f\left(t, x(t), \int_{t_0}^t k(t, s)x(s)\, ds\right) \tag{34}$$

is a causal operator on $L^2([t_0, t_0 + \delta], \mathbb{R}^n)$, continuous and bounded.
   The fact that W acts on L^2 results from (33). The causality is evident. The continuity is a consequence of a result of M. Krasnoselskii (see, for instance, P. Zabrejko *et al.* [1]). Finally, the boundedness also follows from condition (33). This ends the existence proof for equation (32).

Remark   A global variant of this result will be discussed in a subsequent section of this chapter.

   In concluding this section we shall consider one more application of Corollary 3.4 to the case of delay equations of the (general) form (8), under initial condition (9).
   We shall assume that the map f is acting from

$$[t_0, T] \times L^2([t_0, T], \mathbb{R}^n)$$

into $\mathbb{R}^n$. Moreover, let us also assume that for $x \in L^2([t_0, T], \mathbb{R}^n)$ one has $f(t, x_t) \in L^2([t_0, T], \mathbb{R}^n)$. Of course, we take into account the initial condition (9). It is known that in the case of the initial function space consisting of measurable functions, one has to take into account the usual initial condition $x(t_0) = x^0$. Hence, one associates to (8) initial conditions of the form

$$x_0(s) \in L^2([t_0 - h, t_0], \mathbb{R}^n), \quad x(t_0) = x^0 \in \mathbb{R}^n. \tag{35}$$

The space L^2 is here the underlying space, as opposed to the space C we have dealt with in the preceding section of Chapter 3 (for both solution and initial functions).
   The functional equation equivalent to our problem is

$$x(t) = x^0 + \int_{t_0}^t f(s, x_s)\, ds \tag{36}$$

and the right-hand side is obviously causal, acting from

$$[t_0, T] \times L^2([t_0, T], \mathbb{R}^n)$$

into the space of absolutely continuous functions $AC([t_0, T], \mathbb{R}^n) \subset L^2([t_0, T], \mathbb{R}^n)$. Equation (36) is similar to Equation (10), the main difference consisting in the fact that (36) is now considered in the space $L^2([t_0, T], \mathbb{R}^n)$, instead of the space $C([t_0, T], \mathbb{R}^n)$.

In order to apply Corollary 3.4, we need to secure the continuity of the operator

$$(Vx)(t) = x^0 + \int_{t_0}^t f(s, x_s)\, ds$$

on the space $L^2([t_0, T], \mathbb{R}^n)$, as well as its boundedness. These properties can be assured in various ways. We choose here an approach which also leads to uniqueness of the solution. Namely, let f be such that

$$|f(t, \varphi) - f(t, \psi)| \le \lambda(t)|\varphi - \psi|_2, \tag{37}$$

where $\lambda \in L^2([t_0, T], \mathbb{R}^n)$, and the subscript 2 indicates the norm in $L^2([t_0 - h, t_0], \mathbb{R}^n)$. Then, one obtains for $x, y \in L^2([t_0, T], \mathbb{R}^n)$

$$|(Vx)(t) - (Vy)(t)| = \left| \int_{t_0}^t [f(s, x_s) - f(s, y_s)]\, ds \right|$$

$$\le \int_{t_0}^t \lambda(s)|x_s - y_s|_2\, ds$$

$$\le \left(\int_{t_0}^t \lambda^2(s)\, ds \right)^{1/2} \left(\int_{t_0}^t |x_s - y_s|_2^2\, ds \right)^{1/2}.$$

Since both x and y are equal to $x_0(s)$ on $[t_0 - h, t_0]$, it is obvious that, for $s \in [t_0, T]$, $|x_s - y_s|_2 \le |x - y|_2$, the last norm being that of $L^2([t_0, T], \mathbb{R}^n)$. Hence, from the inequality above we derive for $t \in [t_0, T]$

$$|(Vx)(t) - (Vy)(t)| \le (T - t_0)^{1/2} \left(\int_{t_0}^t \lambda^2(s)\, ds \right)^{1/2} |x - y|_2. \tag{38}$$

From (38), taking into account $\lambda \in L^2([t_0, T], \mathbb{R}^n)$, one derives the continuity of the operator $V: L^2([t_0, T], \mathbb{R}^n) \to C([t_0, T], \mathbb{R}^n)$. Since convergence in C on $[t_0, T]$ implies convergence in L^2 on the same interval, it follows that $V: L^2 \to L^2$ is continuous. The boundedness also follows from (38), keeping in mind that a set which is bounded in C is also bounded in L^2 (just square and then integrate both sides of (38)).

The above discussion shows that the conditions of Corollary 3.4 are satisfied, and therefore local existence is assured for the delay equation (36), with initial data (35).

We have stated above that condition (37) also implies uniqueness of the solution. Indeed, from (38) we obtain the estimate

$$\int_{t_0}^t |(Vx)(u) - (Vy)(u)|^2 du \le (T - t_0) \left(\int_{t_0}^t \lambda^2(s)\, ds \right) \left(\int_{t_0}^t |x_s - y_s|^2\, ds \right).$$

This inequality shows that restricting the problem to an interval $[t_0, t_0 + \delta]$, such that

$$(T - t_0)^{1/2} \left(\int_{t_0}^{t_0 + \delta} \lambda^2(s)\, ds \right)^{1/2} < 1,$$

V becomes a contraction on $L^2([t_0, t_0 + \delta], \mathbb{R}^n)$, which means (according to the Banach principle on contraction mappings on L^2) that the equation $x = Vx$ has a unique solution. Of course, this is a local uniqueness result. It turns out that this property is global, which we shall see in subsequent sections of the chapter.

While local existence theorems may not always satisfy the request of applied scientists, they constitute the first step towards a general theory of existence of solutions. The global existence is usually obtained by combining results of local existence with continuation/prolongation methods. Such topics will be discussed in Section 3.4. Again, the causal property makes such things possible.

3.3 Existence and approximation of solutions by means of the singular perturbation technique

The method of singular perturbation is now widely used in problems related to various classes of functional equations: ordinary differential equations, delay equations, difference equations and partial differential equations. For a comprehensive treatment of the subject, in the case of ordinary differential equations, we refer the reader to R. O'Malley, Jr. [1].

The results given below are also included in the author's book [10], and an encouraging feature consists in the fact that the method provides simultaneously existence of solutions and their approximation (or regularization) for rather general equations with causal operators.

We will associate to the functional equation (1) the singularly perturbed functional differential equation

$$\varepsilon \dot{x}(t) = -x_\varepsilon(t) + (Vx_\varepsilon)(t), \quad t \in [t_0, T], \tag{39}$$

in which $\varepsilon > 0$ is assumed to be a small number.

The term "singular perturbation" is probably inspired by the fact that ε multiplies the first derivative $\dot{x}_\varepsilon(t)$, and when $\varepsilon \to 0$ we face a discontinuity in equation (39).

We will pay attention to the case when the operator V in (1) and in (39) is acting on L^p-spaces. The case when $C([t_0, T], \mathbb{R}^n)$ is the underlying space can be treated similarly (see C. Corduneanu [10]), also taking into account the "fixed initial value property" that has to be imposed on the operator V. On the other hand, the solutions whose existence has been proved in Theorem 3.1 or its Corollary 3.2, are continuous or continuously differentiable functions, the regularizing procedure provided by the method of singular perturbation having less significance than in the case of L^p-solutions.

Let us notice that the functional differential equation (39) can be written in integral form as

$$x_\varepsilon(t) = c \exp\left(-\frac{t - t_0}{\varepsilon}\right) + \varepsilon^{-1} \int_{t_0}^t \exp\left(-\frac{t - s}{\varepsilon}\right) (Vx)(s)\, ds, \tag{40}$$

where $c \in \mathbb{R}^n$ is an arbitrary vector and $\varepsilon > 0$. We shall see that it is immaterial which value we assign to c. The final result (as $\varepsilon \to 0$) will be the same.

If we use the notation given by formula (2.67), then (40) becomes

$$x_\varepsilon(t) = c\varepsilon \phi_\varepsilon(t - t_0) + \int_{t_0}^t \phi_\varepsilon(t - s)(Vx_\varepsilon)(s)\, ds. \tag{41}$$

It is easily seen that $\varepsilon |\phi_\varepsilon(t - t_0)|_p \to 0$ as $\varepsilon \to 0$, $1 \leq p < \infty$, which suggests that instead of (41) one can deal with the simpler equation

$$x_\varepsilon(t) = \int_{t_0}^t \phi_\varepsilon(t - s)(Vx_\varepsilon)(s)\,ds. \tag{42}$$

The operator in the right-hand side of (42) is obviously a causal operator, acting on the space $L^p([t_0, T], \mathbb{R}^n)$. The last assertion is a consequence of property (2.66) of the convolution product. For each $\varepsilon > 0$, $\phi_\varepsilon \in L^1([t_0, T], \mathbb{R}^n)$ and V is a continuous and compact operator on $L^p([t_0, T], \mathbb{R}^n)$. Hence, the convolution product in the right-hand side of (42) is a continuous and compact operator on $L^p([t_0, T], \mathbb{R}^n)$. Indeed, V is compact by assumption and, therefore, it takes bounded sets in L^p into relatively compact sets. On the other hand, the convolution is continuous and takes relatively compact sets into relatively compact ones.

To summarize the discussion above, in the right-hand side of (42) we have an L^p-operator which is causal, continuous and compact. In order to apply Schauder's fixed point theorem in L^p, we need to construct a ball in L^p, of fixed radius with respect to ε, such that the ball is taken into itself by the operator in the right-hand side of equation (42). We shall denote by B_r the ball

$$B_r = \{x;\ x \in L^p([t_0, T], \mathbb{R}^n),\ |x|_p \leq r\},$$

for some $r > 0$, and see whether we can achieve

$$T_\varepsilon B_r \subset B_r, \quad \varepsilon > 0, \tag{43}$$

where

$$(T_\varepsilon x)(t) = \int_{t_0}^t \phi_\varepsilon(t - s)(Vx)(s)\,ds.$$

We shall see that an inclusion like (43) is always possible, if the inclusion

$$\int_{t_0}^T |(Vx)(s)|^p\,ds \leq r^p, \quad x \in B_r, \tag{44}$$

is valid. Indeed, from the definition of the operator T_ε, one obtains for $t \in [t_0, T]$ the inequality

$$\int_{t_0}^t |(T_\varepsilon x)(s)|^p\,ds \leq \left[1 - \exp\left(-\frac{T - t_0}{\varepsilon}\right)\right]^p \int_{t_0}^t |(Vx)(s)|^p\,ds \tag{45}$$

assuming also $t_0 \geq 0$, which is not a material restriction. Therefore, (44) implies

$$\int_{t_0}^t |(T_\varepsilon x)(s)|^p\,ds < \int_{t_0}^t |(Vx)(s)|^p\,ds, \tag{46}$$

for any $t \in (t_0, T]$ and $\varepsilon > 0$. Hence, the inclusion (43) is valid any time (44) is valid.

While (44) may not be true in general, we can always find $\delta > 0$, $t_0 + \delta \leq T$, such that

$$\int_{t_0}^{t_0+\delta} |(Vx)(s)|^p\,ds \leq r^p, \quad x \in B_r. \tag{47}$$

This possibility has been discussed in the proof of Theorem 3.3. We have shown that for conveniently chosen $\delta > 0$, (47) holds. This fact suggests that instead of working with the space $L^p([t_0, T], \mathbb{R}^n)$, we should restrict our considerations to the "smaller" space $L^p([t_0, t_0 + \delta], \mathbb{R}^n)$. Then, Schauder's fixed point theorem is applicable, which means every equation (42) has a solution $x_\varepsilon(t)$, $t \in [t_0, t_0 + \delta]$, such that $x_\varepsilon \in L^p([t_0, t_0 + \delta], \mathbb{R}^n)$, and $|x_\varepsilon|_p \leq r$. The existence is also assured for the equation $x(t) = (Vx)(t)$, because of (47).

Let us consider again equation (42), and notice that the compactness of the set $\{Vx_\varepsilon\}$, with $0 < \varepsilon < \varepsilon_0$, $\varepsilon_0 > 0$ fixed, implies the existence of a sequence $\{\varepsilon_m\}$, $\varepsilon_m \to 0$, such that $Vx_{\varepsilon_m} \to u$, $u \in L^p([t_0, t_0 + \delta], \mathbb{R}^n)$. We now claim that the sequence $\{x_{\varepsilon_m}\}$ converges in $L^p([t_0, t_0 + \delta], \mathbb{R}^n)$ to some x, with $Vx = u$. Indeed, from (42) one derives

$$x_{\varepsilon_m}(t) = \int_{t_0}^t \phi_{\varepsilon_m}(t - s)[(Vx_{\varepsilon_m})(s) - u(s)]\, ds + \int_{t_0}^t \phi_{\varepsilon_m}(t - s)u(s)\, ds.$$

As $m \to \infty$, the first integral above tends to zero, since $V\varepsilon_m \to u$ in L^p (one again applies the inequality (2.65)). The second integral above tends to u, in L^p, according to the properties of the approximate unit with respect to the convolution product (see Chapter 2). But the convergence of x_{ε_m} to u, combined with $Vx_\varepsilon \to u$ and the continuity of V, leads to $x = Vx$ $(= u)$.

The above discussion leads us to the conclusion that the singular perturbation approach is legitimate, under the condition specified above. More precisely, the following result can be stated:

Theorem 3.5 *Consider the functional equation* (1), *with the causal operator V on $L^p([t_0, T], \mathbb{R}^n)$, $1 \leq p < \infty$. If V is continuous and compact on L^p, then the singularly perturbed functional differential equation* (39) *has an absolutely continuous solution on some interval $[t_0, t_0 + \delta]$, $t_0 + \delta \leq T$, such that $x_\varepsilon(t_0) = \theta$. Moreover, there exist sequences $\{\varepsilon_m\}$, $\varepsilon_m \to 0$, such that $x_{\varepsilon_m}(t) \to x(t)$ in $L^p([t_0, t_0 + \delta], \mathbb{R}^n)$, where $x(t)$ is a solution of* (1), $x(t) \in L^p([t_0, t_0 + \delta], \mathbb{R}^n)$.

The solutions of (39) can therefore be regarded as "regularized" solutions of (1) in $L^p([t_0, t_0 + \delta], \mathbb{R}^n)$, for small ε.

Corollary 3.6 *Consider the functional differential equation* (2), *and assume the operator V satisfies the hypotheses of Corollary 3.4. Then, the singular perturbation scheme described in Theorem 3.5, applied to the equation*

$$x(t) = \int_{t_0}^t (Vx)(s)\, ds \tag{48}$$

is legitimate. In other words, the solutions $x_\varepsilon(t)$ of the equations

$$\dot{x}_\varepsilon(t) = -x_\varepsilon(t) + \int_{t_0}^t (Vx_\varepsilon)(s)\, ds \tag{49}$$

form a compact set in $L^p([t_0, t_0 + \delta], \mathbb{R}^n)$, $(\varepsilon > 0)$.

The proof is straightforward and is left to the reader. The approximate solutions $x_\varepsilon(t)$ are continuously differentiable and their second derivatives $\ddot{x}_\varepsilon(t)$ are absolutely continuous on $[t_0, t_0 + \delta]$.

It can easily be seen that one could start with the second-order functional differential equation

$$\ddot{x}_\varepsilon(t) = -\dot{x}_\varepsilon(t) + (Vx_\varepsilon)(t), \tag{50}$$

and choose the solution vanishing at t_0 with its first derivative. Then, the equivalence of (49) and (50) becomes transparent.

There are relatively few sources on the method of singular perturbation for functional equations with causal operators, as opposed to classical cases (integral, integrodifferential equations or delay equations).

For instance, if one considers "first kind" equations with causal operators

$$(Vx)(t) = f(t), \tag{51}$$

which are known not to possess solutions in the classical setting (unless f belongs to the range of V), it would be interesting (in the case of existence of solutions) to investigate the singularly perturbed equations

$$\varepsilon x_\varepsilon(t) + (Vx_\varepsilon)(t) = f(t),$$

or

$$\varepsilon \dot{x}_\varepsilon(t) + (Vx_\varepsilon)(t) = f(t),$$

for which one has existence results, as well as their connection with (51).

3.4 Global existence results for functional equations with causal operators

We shall start this section with some elementary considerations on the Schauder fixed point theorem, which can then be easily applied to the functional equation (1), $x(t) = (Vx)(t)$, in order to obtain some global existence results.

First, let $T : B \to B$, B a Banach space, be continuous and taking bounded sets into relatively compact sets. If T is bounded on B, i.e., there exists $M > 0$ such that $|Tx| \le M$ for all $x \in B$, then T has a fixed point $\bar{x}$, with $|\bar{x}| \le M$.

Indeed, the ball of radius M centered at the origin $\theta \in B$ is taken into itself, and TB is relatively compact. Hence, there exists a fixed point $\bar{x}$ for T. Obviously, $|\bar{x}| \le M$.

Second, since the boundedness of T appears as a very strong restriction, one can think of imposing some condition on the growth of T, as $|x| \to \infty$.

Let us denote for $r > 0$

$$\phi(r) = \sup\{|Tx|;\ x \in B,\ |x| \le r\}, \tag{52}$$

which makes sense because T takes bounded sets into relatively compact sets (hence, bounded). The growth condition

$$\limsup_{r\to\infty} [\phi(r)/r] < 1 \tag{53}$$

will assure the fact that T takes a ball of sufficiently large radius, centered at $\theta \in B$, into itself. Since T is assumed continuous and compact, Schauder's theorem is applicable if one chooses the ball of radius R, such that

$$\phi(R)/R < k, \tag{54}$$

with k satisfying

$$\limsup_{r\to\infty} [\phi(r)/r] < k < 1. \tag{55}$$

The existence of such R follows easily from the definition of lim sup.

But condition (54), which can also be written as $\phi(R) < kR$, shows that the ball of radius R, centered at θ, is taken by the operator T (see (52)) into a subset of this ball (radius $kR < R$). Therefore, there exists a fixed point of T, say $\tilde{x}$, such that $|\tilde{x}| < R$. The case of bounded TB is obviously covered by assumption (53), because $\phi(r) = \text{const.} > 0$ for large values of r.

The application of the above statements immediately leads to some elementary global results.

Theorem 3.7   *Consider equation* (1) *in which* $V : E \to E$ *is a continuous and compact causal operator. Assume that condition* (53) *holds, where the norm in* (52) *is that of the space* $E = L^p([t_0, T], \mathbb{R}^n)$, $1 \leq p < \infty$, *or of* $E = C([t_0, T], \mathbb{R}^n)$. *Then there exists a global solution of* (1), *defined on the whole interval* $[t_0, T]$, *belonging to the space* E.

The proof is an immediate consequence of the above discussion.

Corollary 3.8   *If condition* (53) *is replaced by the boundedness of the set* $VE \subset E$, *in the norm of* E, *then the conclusion of Theorem* 3.7 *remains valid.*

See the independent proof in the Remark to Corollary 3.8.

Let us now consider an application of Corollary 3.8 to the equation

$$x(t) = f(t) + \int_{t_0}^{t} k(t, s, x(s))\, ds, \quad t \in [t_0, T], \tag{56}$$

under conditions leading to the (global) existence of a continuous solution. Namely, assume that the following two conditions are satisfied:

1   $f \in C([t_0, T], \mathbb{R}^n)$;
2   $k(t, s, x)$ is a continuous function from $\Delta \times \mathbb{R}^n$ into $\mathbb{R}^n$, $\Delta = \{(t, s); \ t_0 \leq s \leq t \leq T\}$, and is bounded: $|k(t, s, x)| \leq M < \infty$.

We have to show that the operator

$$(Tx)(t) = f(t) + \int_{t_0}^{t} k(t, s, x(s))\, ds \tag{57}$$

is continuous and compact on $C([t_0, T], \mathbb{R}^n)$.

The causality of T is obvious from its definition. The boundedness of TC in C is a consequence of the estimate

$$\sup |(Tx)(t)| \leq \sup |f(t)| + M(T - t_0), \quad t \in [t_0, T].$$

The continuity of the operator T on C is a consequence of the uniform continuity of $k(t, s, x)$ on any set $\Delta \times B$, with B bounded in $\mathbb{R}^n$. Indeed, if $x^{(m)} \to x$ uniformly on $[t_0, T]$, then $\{x^{(m)}\}_{m \geq 1}$ is bounded. We can write $|x^{(m)}(t)| \leq A$, $m \geq 1$, $t \in [t_0, T]$, for some $A > 0$. We shall rely on the uniform continuity of $k(t, s, x)$ on $\Delta \times B$, where B is the ball of radius A, centered at the origin $\theta \in C([t_0, T], \mathbb{R}^n)$.

Since

$$|(Tx^{(m)})(t) - (Tx)(t)| \leq \int_{t_0}^{t} |k(t, s, x^{(m)}(s)) - k(t, s, x(s))| \, ds,$$

and $k(t, s, x)$ is uniformly continuous in $\Delta \times B$, we obtain from the above inequality the continuity of T.

The compactness of T on $C([t_0, T], \mathbb{R}^n)$ is a consequence of the following inequality:

$$|(Tx)(t) - (Tx)(u)|$$

$$\leq |f(t) - f(u)| + \left| \int_u^t |k(t, s, x(s))| \, ds \right| + \int_{t_0}^{T} |k(t, s, x(s)) - k(u, s, x(s))| \, ds.$$

The first term in the right-hand side can be made arbitrarily small for $|t - u|$ small enough, due to the uniform continuity of f on $[t_0, T]$. The second term is dominated by $M|t - u|$, where $M = \sup |k(t, s, x)|$ on $\Delta \times \mathbb{R}^n$. Finally, the third term can be made as small as we want, due to the uniform continuity of k on $\Delta \times B$, where B denotes any bounded set in $\mathbb{R}^n$.

Therefore, under conditions 1 and 2 stated above, *there exists a solution of the equation* (56) *in* $C([t_0, T], \mathbb{R}^n)$.

Remark   The proof above can be easily adapted to cover the case in which boundedness of k is replaced by the more general condition (53). Indeed, one must first obtain $R > 0$, such that (55) is valid. Then one can proceed as above, restricting our considerations on T to the ball of radius R, centered at the origin of $C([t_0, t_0 + \delta], \mathbb{R}^n)$. In this case, $M = \sup |k(t, s, x)|$ on $\Delta \times B_R$, where B_R is the ball of radius R, centered at the origin.

We shall now approach the existence of a global solution for equation (28), choosing the space $L^2([t_0, T], \mathbb{R}^n)$ as the underlying space. This case is similar to the more general case $1 < p < \infty$.

The following assumptions will be made on f, k and G in (28):

1   $f \in L^2([t_0, T], \mathbb{R}^n)$;
2   $k(t, s) = (k_{ij}(t, s))_{n \times n}$ is measurable on Δ, and such that

$$\int_{t_0}^{T} \left(\int_{t_0}^{t} |k(t, s)|^2 \, ds \right) = M^2 < \infty; \tag{58}$$

3   $g(t, x)$ is defined on $[t_0, T] \times \mathbb{R}^n$, satisfies the Carathéodory condition, and

$$|g(t, s)| \leq g_0(t) + c|x|, \tag{59}$$

with $g_0 \in L^2$ and $c > 0$.

Then, equation (28) has a solution $x(t) \in L^2([t_0, t_0 + \delta], \mathbb{R}^n)$, provided c is small enough.

The proof is a consequence of Theorem 3.7, also taking into account that condition (58) assures continuity and compactness of the linear integral operator (31)

$$x(t) \to \int_{t_0}^{t} k(t, s) x(s) \, ds$$

on $L^2([t_0, T], \mathbb{R}^n)$, while condition (59) provides for the continuity (also on L^2) of the operator

$$x(t) \rightarrow g(t, x(t)).$$

All it remains to be proven is the fact that condition (53) is satisfied, provided $c > 0$ is chosen small enough. Considering the operator U defined by (30), one can easily check that it maps $L^2([t_0, T], \mathbb{R}^n)$ into itself. Moreover,

$$|Ux|_2 \leq |f|_2 + \left| \int_{t_0}^t |k(t, s)| g_0(s) ds \right|_2 + c \left| \int_{t_0}^t |k(t, s)| |x(s)| ds \right|_2 \leq K + cM|x|_2.$$

These inequalities show that $\phi(r) \leq K + cMr$, where $\phi(r)$ is defined by (52). Therefore, condition (53) is satisfied if we admit that $cM < 1$. This ends the proof of the existence of an $L^2([t_0, T], \mathbb{R}^n)$ solution of the Hammerstein–Volterra integral equation (28), under conditions 1–3 above.

We will return now to the functional differential equations of the form (2), $\dot{x}(t) = (Vx)(t)$, in view of providing some global existence results. We shall make assumptions on V of such a nature that equation (2) with initial condition $x(t_0) = x^0$ be equivalent to equation (3), i.e.,

$$x(t) = x^0 + \int_{t_0}^t (Vx)(s) \, ds.$$

The result concerning the existence of solutions is derived from Schauder's fixed point theorem. Historically, this result should be related to the names of C. Carathéodory, M. Hukuhara and E.J. McShane [1].

Theorem 3.9 *Let us consider equation (2), with initial condition* $x(t_0) = x^0$, *under the following assumptions:*

1 *V is a causal continuous operator from $C([t_0, T], \mathbb{R}^n)$ into $L^1([t_0, T], \mathbb{R}^n)$;*
2 *there exist two real valued functions $A(t)$ and $B(t)$, defined on $[t_0, T]$, with $A(t)$ continuous and positive and $B(t)$ integrable, such that*

$$\int_{t_0}^t B(s) \, ds \leq A(t) - A(t_0), \quad t \in [t_0, T], \tag{60}$$

while $x(t) \in C([t_0, T], \mathbb{R}^n)$ and

$$|x(t)| \leq A(t), \quad t \in [t_0, T], \tag{61}$$

imply

$$|(Vx)(t)| \leq B(t), \quad a.e. \text{ on } [t_0, T]. \tag{62}$$

Then, there exists a solution $x(t) \in C([t_0, T], \mathbb{R}^n)$ of our problem, provided $|x^0| \leq A(0)$. This solution satisfies the estimate (61).

The proof of Theorem 3.9 can be routinely conducted by means of Schauder's fixed point theorem. Denote by K, $K \subset C$, the set of those functions satisfying (61). This is a convex closed set in C. On behalf of conditions (60) and (62), one can see that the operator W, defined by (5), and $|x^0| \leq A(0)$, takes K into itself: $WK \subset K$. Because of assumption 1 and condition (62), W is continuous on K. The only property that remains to be proved is the compactness of W on K. Or, equivalently, the compactness of the set WK in $C([t_0, T], \mathbb{R}^n)$. We notice that WK consists of absolutely continuous functions on $[t_0, T]$, with values in $\mathbb{R}^n$. Since we have chosen the space of continuous functions as the underlying space, this space also contains the absolutely continuous functions. The solution will be in the space $AC([t_0, T], \mathbb{R}^n)$, because one can differentiate both sides of (3), a.e. on $[t_0, T]$.

For compactness in C we have the Arzelà–Ascoli criterion, which we can easily apply. Indeed, the equicontinuity of the functions in WK follows from (62), taking into account that for $t, u \in [t_0, T]$ we have

$$|(Wx)(t) - (Wx)(u)| \leq \left| \int_u^t B(s)\, ds \right| \tag{63}$$

for each $x \in K$. Due to the integrability of $B(t)$ on $[t_0, T]$, (63) proves the assertion.

Therefore, the operator W satisfies all the conditions required by Schauder's fixed point theorem: $WK \subset K \subset C([t_0, T], \mathbb{R}^n)$, with WK bounded and equicontinuous in C. Hence, W has a fixed point in the set K, which ends the proof.

Remark   Since $L^p([t_0, T], \mathbb{R}^n) \subset L^1([t_0, T], \mathbb{R}^n)$ for $p > 1$, our assumption that V is acting from C into L^1 appears to be the less restrictive in this context. Of course, one could obtain a similar result when L^1 is replaced by L^p, with $p > 1$.

The existence result proven above can be applied to many particular equations of the form (2), by conveniently choosing the operator V from C into L^1. We will consider below some examples.

Assume we are given the linear system of ordinary differential equations in $\mathbb{R}^n$,

$$\dot{x}(t) = L(t)x(t) + f(t), \quad t \in [t_0, T], \tag{64}$$

where $L(t) = (\ell_{ij}(t))_{n \times n}$ is a matrix with L^1 entries, while $f \in L^1([t_0, T], \mathbb{R}^n)$. The usual initial condition $x(t_0) = x^0 \in \mathbb{R}^n$ is associated to (64). In this case, the operator V, from C into L^1, is given by

$$(Vx)(t) = (L)(t)x(t) + f(t), \tag{65}$$

while

$$(Wx)(t) = x^0 + \int_{t_0}^t f(s)\, ds + \int_{t_0}^t L(s)x(s)\, ds \tag{66}$$

acts from C into itself. One can choose

$$B(t) = |L(t)|A(t) + |f(t)|, \tag{67}$$

which is obviously integrable on $[t_0, T]$. Then, condition (60) becomes

$$\int_{t_0}^{t} (|L(s)|A(s) + |f(s)|)\, ds \leq A(t) - A(t_0),\tag{68}$$

which represents an integral inequality for $A(t), t \in [t_0, T]$. It can be rewritten as

$$A(t) \geq A(t_0) + \int_{t_0}^{t} |f(s)|\, ds + \int_{t_0}^{t} |L(s)|A(s)\, ds.\tag{69}$$

The inequality (69) is implied by the stronger inequality

$$A(t) \geq A(t_0) + \int_{t_0}^{T} |f(s)|\, ds + \int_{t_0}^{t} |L(s)|A(s)\, ds$$

which is of the form

$$A(t) \geq C + \int_{t_0}^{t} |L(s)|A(s)\, ds,\tag{70}$$

with

$$C = A(t_0) + \int_{t_0}^{T} |f(s)|\, ds.\tag{71}$$

Since $A(t_0) > 0$ can be chosen arbitrarily, it follows that C is a positive constant larger than the L^1-norm of f on $[t_0, T]$.

It is an elementary exercise to prove that a solution for (70) is the function

$$A(t) = C \exp \left\{ \int_{t_0}^{t} |L(s)|\, ds \right\}.$$

This ends the proof of the global existence on $[t_0, T]$ for the linear system (64).

The uniqueness of the solution of (64), under initial condition $x(t_0) = x^0$, can also be proven by means of integral inequalities. This topic will be discussed in subsequent sections of this chapter and in Chapter 4.

Let us point out the fact that the general result on global existence for the solutions of equation (2), given in Theorem 3.9, is applicable to other types of functional differential equations, such as

$$\dot{x}(t) = f(t, x(t), x(t - h_1), \ldots, x(t - h_m)),$$

with functional initial condition of the form (9). The linear case is described by the equation

$$\dot{x}(t) = L(t)x(t) + \sum_{k=1}^{m} L_k(t)x(t - h_k) + f(t),$$

under suitable conditions (similar to those related to equation (64)). We leave to the reader the task of formulating the conditions assuring the existence of solutions.

We shall provide one more application of Theorem 3.9 to integrodifferential equations of the form (34). In this case, the operator V is given by

$$(Vx)(t) = f\left(t, x(t), \int_{t_0}^{t} k(t, s) x(s)\, ds\right), \tag{72}$$

while the initial condition $x(t_0) = x^0$ is associated to equation (34). In order to satisfy the conditions required by Theorem 3.9, we shall make the following assumptions on the operator V, defined by (72):

1 $f : [t_0, T] \times \mathbb{R}^n \times \mathbb{R}^n \to \mathbb{R}^n$ is a Carathéodory function, i.e., is continuous in (x, y) for almost all $t \in [t_0, T]$, and measurable in t, for fixed $(x, y) \in \mathbb{R}^n \times \mathbb{R}^n$;
2 $k(t, s)$ is a matrix valued kernel of type $n \times n$, continuous on $\Delta = \{(t, s); t_0 \leq s \leq t \leq T\}$.

These two conditions, together with the growth condition on f

$$|f(t, x, y)| \leq \alpha(t)|x| + \beta(t)|y| + \gamma(t), \tag{73}$$

where $\alpha(t)$, $\beta(t)$ and $\gamma(t)$ are nonnegative in $L^1([t_0, T], \mathbb{R}^n)$, assure the fact that V is acting from $C([t_0, T], \mathbb{R}^n)$ into $L^1([t_0, T], \mathbb{R}^n)$.

It is obvious that V, defined by (72), is a causal operator. The continuity follows on behalf of (73), by means of the Lebesgue dominated convergence criterion.

It remains to construct the functions $A(t)$ and $B(t)$, with the properties described in the statement of Theorem 3.9.

We will assume that $A(t)$ is a nondecreasing function on $[t_0, T]$. Then, we can obviously choose as $B(t)$ the function

$$B(t) = \alpha(t)A(t) + \beta(t)A(t)\int_{t_0}^{t} |k(t, s)|\, ds + \gamma(t). \tag{74}$$

In order to satisfy the condition (60), one must have

$$A(t) - A(t_0) \geq \int_{t_0}^{t}\left[\alpha(s) + \beta(s)\int_{t_0}^{s} |k(s, u)|du\right] A(s)\, ds + \int_{t_0}^{t}\gamma(s)\, ds.$$

This inequality has the same form as inequality (69), which we have investigated and found a continuous (actually, absolutely continuous) solution which is nondecreasing on $[t_0, T]$.

Therefore, under conditions 1, 2 and equation (73) above, the integrodifferential equation (34) has a solution with $x(t_0) = x^0 \in \mathbb{R}^n$.

We now return to the basic functional equation (1), $x(t) = (Vx)(t), t \in [t_0, T]$, in order to prove a theorem of global existence on the (semi-open) interval $[t_0, T)$. The case $T = +\infty$ is not excluded. We will choose the space $L^2_{\text{loc}}([t_0, T), \mathbb{R}^n)$ as the underlying space for our problem. Instead of Schauder's fixed point theorem, we shall use now the Tychonoff fixed point theorem in the Fréchet space $L^2_{\text{loc}}([t_0, T), \mathbb{R}^n)$. See Chapter 2, Section 2.4, for the statement of Tychonoff's theorem.

Theorem 3.10 *Consider the functional equation* (1), *and assume V is an operator on the space* $L^2_{\text{loc}}([t_0, T), \mathbb{R}^n)$, *satisfying the conditions of causality, continuity and compactness.*

Furthermore, assume there exist two functions $A, B : [t_0, T) \to [0, \infty)$, with A continuous and positive, and B locally integrable on $[t_0, T)$, such that

$$x \in L^2_{\text{loc}} \quad \text{and} \quad \int_{t_0}^t |x(s)|^2 \, ds \le A(t), \quad t \in [t_0, T), \tag{75}$$

imply

$$|(Vx)(t)|^2 \le B(t), \quad a.e. \text{ on } [t_0, T), \tag{76}$$

while

$$\int_{t_0}^t B(s) \, ds \le A(t), \quad t \in [t_0, T). \tag{77}$$

Then, there exists a solution $x \in L^2_{\text{loc}}([t_0, T), \mathbb{R}^n)$ of the equation (1), satisfying (75).

Proof In the space $L^2_{\text{loc}}([t_0, T), \mathbb{R}^n)$ we consider the closed convex set

$$K = \left\{ x \in L^2_{\text{loc}}([t_0, T), R^n); \int_{t_0}^t |x(s)|^2 \, ds \le A(t), \ t \in [t_0, T) \right\}, \tag{78}$$

which satisfies

$$VK \subset K. \tag{79}$$

Indeed, if $x \in K$, then inequality (76) holds, and taking (77) into account one obtains (79).

By assumption, V is continuous on L^2_{loc} and compact. Therefore, Tychonoff's theorem applies, yielding the global existence of a solution to $x(t) = (Vx)(t)$. Obviously, this solution will satisfy (75).

Remark A local existence result is obtained for equation (1) if we drop the condition (77) from the list of our assumptions. Indeed, (77) will be valid on a small interval $[t_0, t_0 + \delta]$, if A and B exist and satisfy (75) and (76).

As an application of Theorem 3.10, we will now consider the classical linear equation of Volterra type

$$x(t) = f(t) + \int_{t_0}^t k(t, s)x(s) \, ds, \quad t \in [t_0, T), \tag{80}$$

where $f \in L^2_{\text{loc}}([t_0, T), \mathbb{R}^n)$ and $k(t, s)$ is a matrix valued kernel, of type $n \times n$, whose entries are in $L^2_{\text{loc}}(\Delta, R)$, $\Delta = \{(t, s); t_0 \le s \le t < T\}$. This means that for every $t_1 < T$ one has

$$\int_{t_0}^{t_1} \int_{t_0}^t |k(t, s)|^2 \, ds \, dt \le M(t_1) < \infty, \tag{81}$$

for some positive $M(t), t \in [t_0, T)$.

Corollary 3.11 *The linear Volterra equation (80), with $f \in L^2_{\text{loc}}([t_0, T), \mathbb{R}^n)$ and k satisfying (81), has a (global) solution which belongs to $L^2_{\text{loc}}([t_0, T), \mathbb{R}^n)$.*

Proof   In the case of equation (80), the operator V is given by

$$(Vx)(t) = f(t) + \int_{t_0}^{t} k(t, s)x(s)\, ds, \quad t \in [t_0, T), \tag{82}$$

which acts on $L^2_{\text{loc}}([t_0, T), \mathbb{R}^n)$ and is obviously a causal operator. If one chooses arbitrarily $t_1 < T, t_1 > t_0$, then condition (81) implies that Vx restricted to $[t_0, t_1]$ is in $L^2([t_0, t_1], \mathbb{R}^n)$. This is what we mean by $Vx \in L^2_{\text{loc}}([t_0, T), \mathbb{R}^n)$. Actually, condition (81) assures the continuity of V on L^2_{loc}, due to the inequality

$$\int_{t_0}^{t_1} |(Vx)(t) - (Vy)(t)|^2\, dt \leq \left(\int_{t_0}^{t_1} \int_{t_0}^{t} |k(t,s)|^2\, ds\, dt \right) \left(\int_{t_0}^{t_1} |x(s) - y(s)|^2\, ds \right).$$

The compactness is also implied by the condition (81) on the kernel. The Riesz compactness criterion (see Chapter 2, Section 2.2) applies. From the inequality

$$\int_{t_0}^{t_1} |(Vx)(t + h) - (Vx)(t)|^2\, dt$$

$$\leq 3 \int_{t_0}^{t_1} |f(t + h) - f(t)|^2\, dt + 3A(t_1 + h) \int_{t_0}^{t_1} \int_{t}^{t+h} |k(t + h, s)|^2\, ds\, dt$$

$$+ 3 \left(\int_{t_0}^{t_1} |x(s)|^2\, ds \right) \left(\int_{t_0}^{t_1} \int_{t_0}^{t} |k(t + h, s) - k(t, s)|^2\, ds\, dt \right),$$

which is easily obtained from (82), taking into account the elementary inequality $(a+b+c)^2 \leq 3(a^2 + b^2 + c^2)$, one derives $(h > 0)$

$$\lim_{h \to 0} \int_{0}^{t_1} |(Vx)(t + h) - (Vx)(t)|^2\, dt = 0, \tag{83}$$

uniformly with respect to x in any bounded set of $L^2([t_0, t_1], \mathbb{R}^n)$. The first and third term tend to zero as $h \to 0$ because the translation operator is continuous on any L^2-space. The middle term also tends to zero with h, due to assumption (81). These considerations show that V is a compact operator on $L^2_{\text{loc}}([t_0, T), \mathbb{R}^n)$, thus ending the proof of Corollary 3.11.

Remark   As pointed out in the case of differential equation (64), the uniqueness for the integral equation (80) will follow from results we shall prove in subsequent sections.

We will provide a proof for the uniqueness of the solution of (80), by using an argument we borrow from Tricomi's book [1]. In order to prove uniqueness for (80), it is obviously sufficient to prove that the associated homogeneous equation

$$y(t) = \int_{t_0}^{t} k(t, s)y(s)\, ds, \quad t \in [t_0, T), \tag{84}$$

has only the solution $y(t) = 0$, a.e. on $[t_0, T)$. Since (84) yields

$$|y(t)| \leq \int_{t_0}^{t} |k(t, s)||y(s)|\, ds, \quad t \in [t_0, T), \tag{85}$$

it will be sufficient to prove that (85) implies $y(t) = 0$ a.e. on $[t_0, T)$. Let us denote

$$m(t) = \int_{t_0}^{t} |k(t, s)|^2 \, ds, \quad t \in [t_0, T). \tag{86}$$

The function $m(t)$ is integrable on each interval $[t_0, t_1] \subset [t_0, T)$, as seen from condition (81). From (85) we derive by applying the Schwarz inequality

$$|y(t)|^2 \le m(t) \int_{t_0}^{t} |y(s)|^2 \, ds, \quad t \in [t_0, t_1], \tag{87}$$

with $t_1 < T$ arbitrarily chosen. From (87) one derives for $t \in [t_0, t_1]$

$$|y(t)|^2 \le K m(t), \quad K = \int_{t_0}^{t_1} |y(s)|^2 \, ds. \tag{88}$$

From (88) we obtain

$$\int_{t_0}^{t} |y(s)|^2 \, ds \le K \int_{t_0}^{t} m(s) \, ds. \tag{89}$$

Applying now the Schwarz inequality to (85) after squaring both sides, and taking (88) into account we obtain

$$
\begin{aligned}
|y(t)|^2 &\le \int_{t_0}^{t} |k(t, s)|^2 \, ds \int_{t_0}^{t} |y(s)|^2 \, ds \\
&\le K m(t) \int_{t_0}^{t} m(s) \, ds \\
&= \frac{K}{2!} \frac{d}{dt} \left(\int_{t_0}^{t} m(s) \, ds \right)^2.
\end{aligned}
\tag{90}
$$

We obtain from (90), by integration

$$\int_{t_0}^{t} |y(s)|^2 \, ds \le \frac{K}{2!} \left(\int_{t_0}^{t} m(s) \, ds \right)^2, \quad t \in [t_0, t_1]. \tag{91}$$

Proceeding in the same way, repeatedly, we get the estimate

$$\int_{t_0}^{t} |y(s)|^2 \, ds \le \frac{K}{k!} \left(\int_{t_0}^{t} m(s) \, ds \right)^k, \tag{92}$$

which reduces to (89) for $k = 1$ and to (91) for $k = 2$.

But (92) implies $\int_{t_0}^{t} |y(s)|^2 \, ds = 0$, $t \in [t_0, t_1]$, which is possible only for $y(t) = 0$ a.e. on $[t_0, t_1]$.

Since $t_1 < T$ is arbitrary, we see that $y(t) = 0$ a.e. on $[t_0, T)$. This proves uniqueness for the L^2-solution of the linear integral equation (80).

An important class of nonlinear functional equations is that of quasilinear equations which can be written in the form

$$y(t) = (Ly)(t) + (Ny)(t), \quad t \in [t_0, T], \tag{93}$$

where L stands for a linear causal operator while N denotes, in general, a nonlinear causal operator.

We shall postpone the investigation of such equations until we learn more about linear operators of causal type. Usually, N is chosen as a "mild" nonlinearity and the investigation of (93) relies in greatest part on the properties of the associated linear equation $y(t) = (Ly)(t) + f(t)$. In Chapter 4 we shall deal with such equations.

3.5 Some global results of existence and uniqueness

In many applied problems, existence only (local or global) does not provide a complete answer. The uniqueness of the solution, usually under some type of auxiliary conditions (initial, boundary value) is a very desirable feature for the applied scientist. It is true that nowadays complex physical problems lead to equations with multiple solutions. But at least in the very classical setting, without uniqueness, which is a reflection of physical determinism, it seems rather difficult to handle and interpret the mathematical apparatus.

Historically, Cauchy was the first to be systematically concerned with existence theorems and he really created a good start. But all his results, concerning ordinary differential equations or first-order partial differential equations, are also uniqueness results.

Let us now consider equation (2)

$$\dot{x}(t) = (Vx)(t), \quad t \in [t_0, T), \quad T \leq \infty,$$

and provide results for both existence and uniqueness of the solution. As usual, one associates with (2) the initial condition $x(t_0) = x^0 \in \mathbb{R}^n$. Then (1), (2) is equivalent to equation (3).

A classical procedure that has been used by Cauchy and many followers (among them, E. Picard has his name solidly related to this method), is the iteration one. When applied to various types of functional equations, it is often called the "method of successive approximations".

Proceeding formally with equation (2), with V acting on a certain function space $E([t_0, T), \mathbb{R}^n)$, one forms the sequence of elements of E by choosing an arbitrary initial element $x^{(0)}(t) \in E$, and then letting

$$x^{(m)}(t) = x^0 + \int_{t_0}^{t} (Vx^{(m-1)})(s) \, ds, \quad t \in [t_0, T), \quad m \geq 1. \tag{94}$$

The sequence $\{x^{(m)}(t); \ m \geq 1\}$ may or may not be convergent in the space E. But in case it does converge, it is easy to prove that $x(t) = \lim x^{(m)}(t)$, as $m \to \infty$, is a solution of (2), satisfying the initial condition $x(t_0) = x^0$, belonging to the space $E([t_0, T), \mathbb{R}^n)$, or to a "smaller" space in $E([t_0, T), \mathbb{R}^n)$.

In the usual case, the method of successive approximations can be used to prove both existence and uniqueness. The general scheme of successive approximations is embodied in the Banach contraction mapping principle, which we have proven in Chapter 2, Section 2.4. In Section 2.3 of Chapter 2 we have used the method of successive approximations in connection with the linear integral equation (2.49) and the invertibility of the linear integral operator given by (2.50).

Assume now that we choose the space $C([t_0, T), \mathbb{R}^n)$ as the underlying space for the study of equation (2). We have seen in Chapter 2, Section 2.2, that $C([t_0, T), \mathbb{R}^n)$ is a Fréchet space. We shall discuss in this section the case when the solution belongs to some $C_g([t_0, T), \mathbb{R}^n)$ spaces, with convenient g. The fact that the solution belongs to C_g provides us with some extra information about the behavior of that solution as $t \to T$. The spaces $C_g([t_0, T), \mathbb{R}^n)$ have been defined and briefly investigated in Chapter 2, Section 2.2, when $T = +\infty$. The fact that this assumption is dropped does not change the basic considerations.

Theorem 3.12 *Let us consider the functional differential equation* (2), *with* $V : C([t_0, T), \mathbb{R}^n) \to C([t_0, T), \mathbb{R}^n)$, *and such that*

$$|(Vx)(t) - (Vy)(t)| \leq \lambda(t) \sup_{t_0 \leq s \leq t} |x(s) - y(s)|, \tag{95}$$

where $\lambda : [t_0, T) \to (0, \infty)$ *is a nondecreasing function.*

Then, equation (2) *has a unique solution* $x(t) \in C([t_0, T), \mathbb{R}^n)$, *with* $x(t_0) = x^0 \in \mathbb{R}^n$. *Moreover,* $x(t) \in C^{(1)}([t_0, T), \mathbb{R}^n)$.

Proof We consider the successive approximations $\{x^{(m)}(t); m \geq 1\}$, as described by (94).

We can start with an arbitrary $x^{(1)}(t) \in C([t_0, T), \mathbb{R}^n)$. One obtains from (94) the following recurrent inequality, valid on $[t_0, T)$ and for $m \geq 2$:

$$|x^{(m+1)}(t) - x^{(m)}(t)| \leq \int_{t_0}^{t} \lambda(s) \sup_{t_0 \leq u \leq s} |x^{(m)}(u) - x^{(m-1)}(u)| \, ds,$$

which implies

$$\sup |x^{(m+1)}(t) - x^{(m)}(t)| \leq \int_{t_0}^{t} \lambda(t) \sup_{t_0 \leq u \leq s} |x^{(m)}(u) - x^{(m-1)}(u)| \, ds.$$

Let us denote for $t \in [t_0, T)$ and $m \geq 2$

$$u_m(t) = \sup_{t_0 < s \leq t} |x^{(m)}(s) - x^{(m-1)}(s)|. \tag{96}$$

Then, we have

$$u_{m+1}(t) \leq \int_{t_0}^{t} \lambda(s) u_m(s) \, ds, \quad t \in [t_0, T), \tag{97}$$

for $m \geq 2$. We will now fix $t_1 < T$, $(t_0 < t_1)$. On $[t_0, t_1]$ we have $|u_2(t)| \leq A_2$, $A_2 > 0$. Therefore,

$$u_3(t) \leq A_2 \int_{t_0}^{t} \lambda(s) \, ds, \quad t \in [t_0, t_1].$$

From the last inequality and (97) one obtains for $m = 3$:

$$u_4(t) \leq A_2 \int_{t_0}^{t} \lambda(s) \int_{t_0}^{s} \lambda(u) \, du = \frac{1}{2!} \left(\int_{t_0}^{t} \lambda(s) \, ds \right)^2.$$

If we repeatedly use (97), we obtain

$$u_{m+1}(t) \leq \frac{1}{(m-1)!} A_2 \left(\int_{t_0}^{t} \lambda(s) \, ds \right)^{m-1} \tag{98}$$

for $m \geq 2$ and $t \in [t_0, t_1]$. But (98) shows that the series

$$x^{(1)}(t) + [x^{(2)}(t) - x^{(1)}(t)] + \cdots + [x^{(m+1)}(t) - x^{(m)}(t)] + \cdots$$

is absolutely and uniformly convergent on $[t_0, t_1]$. The partial sums of the above series are nothing else but the terms of the sequence of successive approximations $\{x^{(m)}(t); \ m \geq 1\}$, which implies that

$$\lim_{m \to \infty} x^{(m)}(t) = x(t) \in C([t_0, T), \mathbb{R}^n),$$

due to the arbitrariness of $t_1 < T$.

Since V is continuous on C, from (94) one derives, as $m \to \infty$,

$$x(t) = x^0 + \int_{t_0}^t (Vx)(s) \, ds, \quad t \in [t_0, T).$$

But this is equivalent to $\dot{x}(t) = (Vx)(t), x(t_0) = x^0$.

Notice that the convergence is that of the (Fréchet) space $C([t_0, T), \mathbb{R}^n)$, which means uniform convergence on each compact interval of $[t_0, T)$.

Uniqueness is obtained by the same method of successive approximations, as follows: let $y(t) \in C([t_0, T), \mathbb{R}^n)$ be a solution of (2), i.e.,

$$y(t) = x^0 + \int_{t_0}^t (Vy)(s) \, ds, \quad t \in [t_0, T).$$

Using (94) we obtain the recurrent inequality

$$|y(t) - x^{(m+1)}(t)| \leq \int_{t_0}^t \lambda(s) \sup_{t_0 \leq u \leq s} |y(u) - x^{(m)}(u)| \, ds,$$

which can be processed in exactly the same way we proceed with (97). One obtains

$$|y(t) - x^{(m+1)}(t)| \leq \frac{A}{m!} \left(\int_{t_0}^t \lambda(s) \, ds \right)^m,$$

where $A = \sup |y(s) - x^{(1)}(s)|, t_0 \leq s \leq t_1 < T$. This proves that

$$\lim_{m \to 0} x^{(m)}(t) = y(t)$$

and since the limit is unique, we obtain $x(t) = y(t)$ on each $[t_0, t_1] \subset [t_0, T)$, i.e., on $[t_0, T)$.

Remark 1   Condition (95) assures the causality of V, its continuity on $C([t_0, T), \mathbb{R}^n)$, as well as the property of taking bounded sets into bounded sets. A bounded set in $C([t_0, T), \mathbb{R}^n)$ consists of maps uniformly bounded on each $[t_0, t_1] \subset [t_0, T)$.

Remark 2   Stating that $\lambda(t)$ is nondecreasing is not a restriction. If $\lambda(t)$ is assumed to be positive only, then one can easily show that it must be nondecreasing. Indeed, if $t_0 < t_1 < t_2$, one must have $\lambda(t_1) \leq \lambda(t_2)$. Otherwise, $\lambda(t_1) < \lambda(t_2)$, and this leads to a contradiction. By extending any $x(t) = x(t_1), t \in [t_1, t_2]$, one obtains a contradiction. Of course, it is assumed that $\lambda(t)$ is the smallest number in (95), for every $t \in (t_0, T)$.

Remark 3   It is desirable to find a C_g space, such that the solution $x(t) \in C_g([t_0, T), \mathbb{R}^n)$. More exactly, what should be the relationship between g and λ such that $x(t) \in C_g$?

If we assume that $(V\theta)(t) = \theta$ on $[t_0, T)$, then we obtain from

$$x(t) = x^0 + \int_{t_0}^t [(Vx)(s) - (V\theta)(s)] \, ds, \quad t \in [t_0, T),$$

$$|x(t)| \le |x^0| + \int_{t_0}^t \lambda(s) \sup_{t_0 \le u \le s} |x(u)| \, ds, \quad t \in [t_0, T),$$

which implies for $u(t) = \sup |x(s)|, t_0 \le s \le t,$

$$u(t) \le |x^0| + \int_{t_0}^t \lambda(s) u(s) \, ds, \quad t \in [t_0, T). \tag{99}$$

This is an integral inequality known as Gronwall's inequality. We shall discuss inequalities in the next section of this chapter. As we shall see, (99) implies

$$u(t) \le |x^0| \exp\left\{ \int_{t_0}^t \lambda(s) \, ds \right\},$$

which means

$$|x(t)| \le \sup_{t_0 \le s \le t} |x(s)| \le |x^0| \exp\left\{ \int_{t_0}^t \lambda(s) \, ds \right\}. \tag{100}$$

From inequality (100) we read $x(t) \in C_g([t_0, T), \mathbb{R}^n)$, with $g(t) = \exp\{\int_{t_0}^t \lambda(s) \, ds\}$.

The case $(V\theta)(t) \ne \theta$ can be discussed by the same method of integral inequalities, and in general, the estimate (100) is not valid. One can obtain similar estimates for the solutions, making convenient assumptions on the growth of $(V\theta)(t)$. We will discuss such examples in the next section.

As an application of Theorem 3.12 we shall deal with the integrodifferential equation

$$\dot{x}(t) = A(t)x(t) + \int_{t_0}^t B(t, s)x(s) \, ds + f(t), \tag{101}$$

where $A(t)$ is a continuous square matrix of type $n \times n$, $B(t, s)$ is also continuous on $\Delta = \{(t, s); \ t_0 \le s \le t < T\}$ and of the same type, while $f : [t_0, T) \to \mathbb{R}^n$ is continuous.

If one denotes by $(Vx)(t)$ the right-hand side of (101), it can easily be seen that (95) holds, with

$$\lambda(t) = |A(t)| + \int_{t_0}^t |B(t, s)| \, ds.$$

We shall consider again the functional differential equation (2), under different conditions from those of Theorem 3.12. Namely, instead of assuming $V : C([t_0, T), \mathbb{R}^n) \to C([t_0, T), \mathbb{R}^n)$, we shall assume $V : C([t_0, T), \mathbb{R}^n) \to L^1_{\text{loc}}([t_0, T), \mathbb{R}^n)$, which will enable us to obtain absolutely continuous solutions.

Let us now formulate the basic assumption on the operator V in equation (2). This will imply causality and continuity of this operator. Namely, we assume that V satisfies the generalized Lipschitz-type condition for $t \in [t_0, T)$,

$$\int_{t_0}^t |(Vx)(s) - (Vy)(s)| \, ds \le \int_{t_0}^t \lambda(s)|x(s) - y(s)| \, ds, \tag{102}$$

where $\lambda(\cdot)$ is a real-valued function on $[t_0, T)$, locally integrable and nonnegative, while $x, y \in C([t_0, T), \mathbb{R}^n)$. Condition (102) is a natural condition (if we want uniqueness!) when $V : C \to L^1_{loc}$.

We could proceed now by successive approximations as in Theorem 3.12. We prefer to use a fixed point argument, which will allow us to find an estimate providing some informtion about the growth of the solution, for $t \to T$.

By the contraction mapping theorem, we shall prove that the operator

$$x(t) \to x^0 + \int_{t_0}^t (Vx)(s) \, ds, \tag{103}$$

under condition (102), has a fixed point in a convenient $C_g([t_0, T), \mathbb{R}^n)$ space. If one denotes by $(Ux)(t)$ the operator in the right-hand side of (103), which obviously takes its values in $C([t_0, T), \mathbb{R}^n)$, then we obtain on behalf of (102) for $t \in [t_0, T)$,

$$|(Ux)(t) - (Uy)(t)| \leq \int_{t_0}^t \lambda(s)|x(s) - y(s)| \, ds. \tag{104}$$

Instead of using the usual supremum norm, we will use a weighted norm, which will lead to the contraction mapping. We will proceed as follows:

$$|(Ux)(t) - (Uy)(t)|$$

$$\leq \int_{t_0}^t \lambda(s) e^{\alpha \int_{t_0}^s \lambda(u) \, du} \left(|x(s) - y(s)| e^{-\alpha \int_{t_0}^s \lambda(u) \, du} \right) ds$$

$$\leq \sup_{t_0 \leq t < T} \left(|x(t) - y(t)| e^{-\alpha \int_{t_0}^t \lambda(u) \, du} \right) \int_{t_0}^t \lambda(s) e^{\alpha \int_{t_0}^s \lambda(u) \, du} \, ds$$

$$\leq \alpha^{-1} e^{\alpha \int_{t_0}^t \lambda(u) \, du} \sup_{t_0 \leq t < T} \left(|x(t) - y(t)| e^{-\alpha \int_{t_0}^t \lambda(u) \, du} \right).$$

In these inequalities we multiply the first and last term by $\exp\{-\alpha \int_{t_0}^t \lambda(u) \, du\}$, and obtain

$$|(Ux)(t) - (Uy)(t)| e^{-\alpha \int_{t_0}^t \lambda(u) \, du}$$

$$\leq \alpha^{-1} \sup_{t_0 \leq t < T} \left(|x(t) - y(t)| e^{-\alpha \int_{t_0}^t \lambda(u) \, du} \right). \tag{105}$$

We have tacitly admitted that x and y are from the function space $C_g([t_0, T), \mathbb{R}^n)$, with

$$g(t) = \exp\left\{ \alpha \int_{t_0}^t \lambda(s) \, ds \right\}, \quad \alpha > 1. \tag{106}$$

Based on (106), it appears from (105) that the operator U is a contraction on C_g:

$$|Ux - Uy|_g \leq \alpha^{-1} |x - y|_g.$$

Still, we have to prove that

$$U C_g \subset C_g, \tag{107}$$

for g given by (106) and U satisfying (104). This may not be true in general, but it is true in case $(V\theta)(t)$ satisfies a certain estimate. More precisely, it suffices to assume that for some $K > 0$

$$|(V\theta)(t)| \leq K\lambda(t) \exp\left\{\alpha \int_{t_0}^{t} \lambda(s)\, ds\right\}. \tag{108}$$

Then, we obtain for Ux the following estimate:

$$|(Ux)(t)| \leq |x^0| + \int_{t_0}^{t} |(Vx)(s) - (V\theta)(s)|\, ds + \int_{t_0}^{t} |(V\theta)(s)|\, ds$$

$$\leq |x^0| + \int_{t_0}^{t} \lambda(s)|x(s)|\, ds + K\int_{t_0}^{t} \lambda(s) \exp\left\{\alpha \int_{t_0}^{s} \lambda(u)\, du\right\} ds.$$

Since $x \in C_g$, one easily obtains from the preceding inequality

$$|(Ux)(t)| \leq (|x^0| + \alpha^{-1}(A_x + K)) \exp\left\{\alpha \int_{t_0}^{t} \lambda(u)\, du\right\},$$

showing that (107) is satisfied.

Therefore, the operator U defined by (103) takes the space $C_g([t_0, T), \mathbb{R}^n)$ into itself and it is a contraction mapping.

The discussion carried out above allows us to state the following result.

Theorem 3.13 *Consider the functional differential equation* (2), *with* $V : C([t_0, T), \mathbb{R}^n) \to L^1_{loc}([t_0, T), \mathbb{R}^n)$ *satisfying* (102) *and* (108).

Then, there exists a unique solution $x \in C_g([t_0, T), \mathbb{R}^n)$, *with* g *given by* (106), *satisfying the initial condition* $x(t_0) = x^0$. *This solution is locally absolutely continuous on* $[t_0, T)$.

Remark By straightforward application of the contraction mapping theorem, one obtains uniqueness only in the space C_g. Actually, uniqueness is assured in the (much larger) space $C([t_0, T), \mathbb{R}^n)$. This follows from the inequality

$$|x(t) - y(t)| \leq \int_{t_0}^{t} \lambda(s)|x(s) - y(s)|\, ds,$$

which we can process in the same way as in the case of proving uniqueness for the result given in Theorem 3.12 (by repeatedly substituting the left-hand side under the integral in the right-hand side), and integrating.

We shall now apply Theorem 3.13 to the integrodifferential equation (32)

$$\dot{x}(t) = f\left(t, x(t), \int_{t_0}^{t} k(t, s)x(s)\, ds\right),$$

where $f : [t_0, T) \times \mathbb{R}^n \times \mathbb{R}^n$ satisfies the Carathéodory conditions (measurable in t, and continuous with respect to second and third argument for almost all t), as well as the generalized Lipschitz-type condition

$$|f(t, x, y) - f(t, \bar{x}, \bar{y})| \leq \lambda(t)[|x - \bar{x}| + |y - \bar{y}|], \tag{109}$$

with $\lambda(t)$ nonnegative on $[t_0, T)$, and locally integrable. Moreover, we assume $k(t, s)$ is a matrix kernel of type $n \times n$, continuous on $\Delta = \{(t, s); \ t_0 \leq s \leq t < T\}$.

From (109) one obtains

$$\left| f\left(t, x(t), \int_{t_0}^t k(t, s)x(s)\,ds\right) - f\left(t, y(t), \int_{t_0}^t k(t, s)y(s)\,ds\right) \right|$$

$$\leq \lambda(t)\left[|x(t) - y(t)| + \int_{t_0}^t |k(t, s)||x(s) - y(s)|\,ds \right].$$

In order to check condition (102) we consider

$$\int_{t_0}^t \left| f\left(s, x(s), \int_{t_0}^s k(s, u)x(u)\,du\right) - f\left(s, y(s), \int_{t_0}^s k(s, u)y(u)\,du\right) \right| ds$$

$$\leq \int_{t_0}^t \lambda(s)\left[|x(s) - y(s)| + \int_{t_0}^s |k(s, u)||x(u) - y(u)|\,du \right] ds.$$

We shall now transform the double integral which appears in the last term above. One has

$$\int_{t_0}^t \lambda(s)\,ds \int_{t_0}^s |k(s, u)||x(u) - y(u)|\,du = \int_{t_0}^s |x(u) - y(u)|\,du \int_{t_0}^u |k(s, u)|\lambda(s)\,ds$$

$$= \int_{t_0}^t k_0(u)|x(u) - y(u)|\,du,$$

with

$$k_0(u) = \int_{t_0}^u |k(s, u)|\lambda(s)\,ds.$$

We can now write the Lipschitz-type condition above in the form (with $V = f$)

$$\int_{t_0}^t |(Vx)(s) - (Vy)(s)|\,ds \leq \int_{t_0}^t [\lambda(s) + k_0(s)]|x(s) - y(s)|\,ds,$$

which is of the form (102), with $\lambda(s) + k_0(s)$ instead of $\lambda(s)$. A condition similar to (108) must also be imposed, in order to assure global existence of the solution to (32), with $x(t_0) = x^0$.

In concluding this section we shall again consider equation (1), $x(t) = (Vx)(t)$, on the closed interval $[t_0, T]$, in view of obtaining an existence and uniqueness result in the space $L^p([t_0, T], \mathbb{R}^n)$.

The following assumptions will be made on the operator V, which acts on the space $L^p([t_0, T], \mathbb{R}^n)$, $1 < p < \infty$.

1   For all $t \in [t_0, T]$ and all $x, y \in L^p$,

$$|(Vx)(t) - (Vy)(t)| \leq \int_{t_0}^t L(t, s)|x(s) - y(s)|\,ds. \tag{110}$$

2 The function $L(t, s)$ is measurable on $\Delta = \{(t, s); t_0 \le s \le t \le T\}$, and such that

$$M(t) = \left(\int_{t_0}^{T} L^q(t, s) \, ds \right)^{p/q} \in L^1([t_0, T], \mathbb{R}^n), \qquad (111)$$

where $p^{-1} + q^{-1} = 1$.

Theorem 3.14 *Under conditions 1 and 2 above, there exists a unqiue solution $x \in L^p([t_0, T], \mathbb{R}^n)$ of the functional equation (1).*

Proof Let $G(t) \in C([t_0, T], \mathbb{R}^n)$ be a positive function (in a sense to be made precise), and $\rho \in (0, 1)$.

Let us now consider the weighted L^p-norm

$$
\begin{aligned}
|Vx - Vy|_{p,G}^p &= \int_{t_0}^{T} |(Vx)(t) - (Vy)(t)|^p G(t) \, dt \\
&\le \int_{t_0}^{T} \left[\int_{t_0}^{t} L(t, s)|x(s) - y(s)| \, ds \right]^p G(t) \, dt \\
&\le \int_{t_0}^{T} \left[\int_{t_0}^{t} |x(s) - y(s)|^p \, ds \right] \left[\int_{t_0}^{t} L^q(t, s) \, ds \right]^{p/q} G(t) \, dt \\
&\le \int_{t_0}^{T} \left[\int_{t_0}^{t} |x(s) - y(s)|^p \, ds \right] M(t)G(t) \, dt \\
&= \int_{t_0}^{T} |x(s) - y(s)|^p \left[\int_{s}^{T} M(t)G(t) \, dt \right] ds.
\end{aligned}
$$

In order to obtain the contraction of V in the norm $|\cdot|_{p,G}$, we will assume now that

$$\int_{s}^{T} M(t)G(t) \, dt \le \rho^p G(s), \qquad s \in [t_0, T]. \qquad (112)$$

Then, from the sequence of inequalities above we obtain

$$|Vx - Vy|_{p,G} \le \rho |x - y|_{p,G}. \qquad (113)$$

The condition (113) shows that V is a contraction on $L^p([t_0, T], \mathbb{R}^n)$ endowed with the $|\cdot|_{p,G}$ norm. Since $G(t)$ is assumed continuous and positive, it is obvious that the usual L^p-norm is equivalent to the $|\cdot|_{p,G}$ norm.

Therefore, the only point to be clarified is the existence of a positive and continuous solution to the integral inequality (112).

In order to produce such a solution, let us introduce the function $\tilde{G}(u) = G(T + t_0 - u)$, $u \in [t_0, T]$. Then (112) becomes

$$\int_{t_0}^{s} N(u)\tilde{G}(u) \, du \le \tilde{G}(s), \qquad s \in [t_0, T], \qquad (114)$$

with $N(u) = \rho^{-p} M(T + t_0 - u)$.

One can easily check that

$$\tilde{G}(s) = \exp\left(\int_{t_0}^s N(u)\,du\right) \tag{115}$$

is a solution to (114), both continuous and positive on $[t_0, T]$.
This ends the proof of Theorem 3.14.

The most direct application of Theorem 3.14 is related to the nonlinear integral equation (6)

$$x(t) = f(t) + \int_{t_0}^t K(t, s, x(s))\,ds, \quad t \in [t_0, T],$$

for which we have proven local existence in the space of continuous functions (Section 1).

The following conditions will assure the direct application of Theorem 3.14, which provides global existence and uniqueness, on the whole interval $[t_0, T]$, in the space $L^p([t_0, T], \mathbb{R}^n)$.

1   $f \in L^p([t_0, T], \mathbb{R}^n)$;
2   $|K(t, s, x) - K(t, s, y)| \leq L(t, s)|x - y|$, for $x, y \in \mathbb{R}^n$, with $L(t, s)$ measurable on $\Delta = \{(t, s); t_0 \leq s \leq t \leq T\}$ and satisfying condition (111).

Another application of Theorem 3.14 is concerned with integrodifferential equations of the form

$$\dot{x}(t) = f(t, x(t)) + \int_{t_0}^t K(t, s, x(s))\,ds, \tag{116}$$

which appear as perturbed versions of the ordinary differential equation $\dot{x}(t) = f(t, x(t))$. More precisely, the perturbation terms include the memory of the system. If one associates with (116) the usual initial condition $x(t_0) = x^0 \in \mathbb{R}^n$, then the problem leads to the functional equation with causal operator

$$x(t) = x^0 + \int_{t_0}^t \left\{ f(s, x(s)) + \int_{t_0}^s K(s, u, x(u))\,du \right\} ds,$$

which can also be written in the form (1), with $(Vx)(t)$ given by

$$(Vx)(t) = x^0 + \int_{t_0}^t \left\{ f(s, x(s)) + \int_{t_0}^s K(s, u, x(u))\,du \right\} ds. \tag{117}$$

One needs to assure $Vx \in L^p$, for $x \in L^p$. Since the right-hand side of (117) is an absolutely continuous function whenever the integrand is in L^1, we should be concerned with the conditions assuring the integrability on $[t_0, T]$ of the right-hand side of (116), when $x \in L^p([t_0, T], \mathbb{R}^n)$. Moreover, a condition of the form (110) has to be valid, with $M(t)$ satisfying (111).

The following set of conditions represents one of the possibilities in meeting the requirements specified above.

1   $f : [t_0, T] \times \mathbb{R}^n \to \mathbb{R}^n$ is measurable and such that

$$|f(t, x) - f(t, y)| \leq g(t)|x - y|, \quad t \in [t_0, T], \tag{118}$$

$x, y \in \mathbb{R}^n$, with $g \in L^q([t_0, T], \mathbb{R}^n)$, while

$$f(t, \theta) \in L^1([t_0, T], \mathbb{R}^n), \tag{119}$$

$(p^{-1} + q^{-1} = 1)$.

2   $K : \Delta \times \mathbb{R}^n \to \mathbb{R}^n$ is measurable, $\Delta = \{(t, s); t_0 \leq s \leq t \leq T\}$, and such that

$$|K(t, s, x) - K(t, s, y)| \leq L(t, s)|x - y|, \tag{120}$$

where $L(t, s)$ satisfies (111). Moreover,

$$K(t, s, \theta) \in L^1(\Delta, \mathbb{R}^n).$$

It is routine to check that conditions (1) and (2) formulated above are sufficient to assure that $(Vx)(t)$ given by (117) is acting on the space $L^p([t_0, T], \mathbb{R}^n)$, and satisfies a generalized Lipschitz condition of the form (110). This suffices to conclude that the integrodifferential equation (116), with initial condition $x(t_0) = x^0 \in \mathbb{R}^n$, has a unique solution in the space $AC([t_0, T], \mathbb{R}^n)$. While it is a tautology to say that the solutions belong to L^p, it is useful to notice that the conditions imposed on f and K, in order to create the L^p-framework, are considerably more general than those necessary to work in the space of continuous functions.

Theorem 3.14 can also be applied to delay-integral equations of the form

$$x(t) = f(t) + \int_{t_0}^{t} K(t, s; x_s) \, ds,$$

with an initial functional condition $x_{t_0} = \varphi$.

3.6 Functional inequalities

It is widely known that differential and integral inequalities play an important role in obtaining uniqueness, continuous dependence of the solution with respect to data, stability results and other basic properties for differential, delay or integral equations.

In this section we shall consider functional inequalities with causal operators, as well as some of their applications to functional equations.

We shall start with the scalar functional inequality

$$y(t) \leq (Vy)(t), \quad t \in [t_0, T), \tag{121}$$

where $V : C([t_0, T), \mathbb{R}) \to C([t_0, T), \mathbb{R})$ is a *continuous* operator.

Extra conditions will be imposed on V, in order to develop a useful theory of such inequalities, and make them applicable to the investigation of functional or functional differential equations with causal operators.

We will assume that the following conditions hold for the operator V appearing in (121).

1   V is causal, continuous and compact on $C([t_0, T), \mathbb{R})$;
2   V is monotonically nondecreasing, i.e., if $x(t), y(t) \in C([t_0, T), \mathbb{R})$ and $x(t) \le y(t)$ on $[t_0, T)$, then $(Vx)(t) \le (Vy)(t)$ on the same interval.

Actually, in what follows, we will consider the operator V only on a subset of the space $C([t_0, T), \mathbb{R})$, namely, on the subset S of C consisting of those functions $x(t)$, such that

$$z_1(t) \le x(t) \le z_2(t), \quad t \in [t_0, T), \tag{122}$$

where $z_1, z_2 \in C$, and satisfy the inequalities

$$z_1(t) \le (Vz_1)(t), \qquad z_2(t) \ge (Vz_2)(t), \tag{123}$$

on the interval $[t_0, T)$. If (123) are satisfied, one calls $z_1(t)$ a *subsolution* and $z_2(t)$ a *supersolution* of the functional equation (1)

$$x(t) = (Vx)(t), \quad t \in [t_0, T),$$

completely defined by the operator V.

The property 2 of V tells us that for any $z(t) \in C([t_0, T), \mathbb{R})$, with $z_1(t) \le z(t) \le z_2(t)$, one has

$$(Vz_1)(t) \le (Vz)(t) \le (Vz_2)(t), \quad t \in [t_0, T),$$

which combined with (123) leads to

$$z_1(t) \le (Vz)(t) \le z_2(t), \quad t \in [t_0, T). \tag{124}$$

Therefore, the set S of functions in C, satisfying (122), is taken into itself by the operator V, $(VS \subset S)$.

But it is easy to check the fact that S is a convex subset of $C([t_0, T), \mathbb{R})$. Indeed, if $x, y \in S$, then $\lambda x + (1 - \lambda)y \in S$ for any $\lambda \subset (0, 1)$, because the inequalities $z_1(t) \le \lambda x(t) + (1 - \lambda)y(t) \le z_2(t), t \in [t_0, T)$, are direct consequences of (122) and of a similar inequality for $y(t)$. Moreover, from $x^{(m)}(t) \in S, m \ge 1, x^{(m)}(t) \to x(t)$, uniformly on any $[t_0, t_1]$, with $t_1 < T$, one obtains (122) for $x(t)$. Hence, the convexity and closedness of S in C are satisfied. Since S is bounded in $C([t_0, T), \mathbb{R})$, its functions being uniformly bounded on each $[t_0, t_1]$, $t_1 < T$, the conditions required by Tychonoff's fixed point theorem in the Fréchet space $C([t_0, T), R)$ are satisfied.

Therefore, equation (1) has a solution in S, say $x(t)$, which means (122) holds true.

As we know from the theory of ordinary differential equations $\dot{x}(t) = f(t, x(t))$, in which case

$$(Vx)(t) = x^0 + \int_{t_0}^t f(s, x(s)) \, ds,$$

the solution is not always unique. We cannot expect uniqueness for equation (1), with V satisfying only conditions 1 and 2 above.

It is easy to see that the set of solutions to (1), belonging to S, is compact in $C([t_0, T), \mathbb{R})$. Indeed, from $x^{(m)}(t) = (Vx^{(m)})(t)$ and $x^{(m)}(t) \to x(t)$ in C, one obtains $x(t) = (Vx)(t)$

as $m \to \infty$, which means that the set of solutions to (1), belonging to VS, is closed in C. Since VS is relatively compact in C, this means that the set of solutions is compact.

Other properties of the solution set could be established, but this is not our main concern here.

A property which is interesting for us in this context is the existence of *extremal solutions* to equation (1), the *minimal solution* $x_m(t)$, and the *maximal solution* $x_M(t)$.

Before we prove the existence of the extremal solutions, such that any solution $x(t)$ of (1) satisfies

$$x_m(t) \le x(t) \le x_M(t), \quad t \in [t_0, T), \tag{125}$$

let us notice that they will allow us to give an answer to the solution of the functional inequality (121), considered at the beginning of this section. The answer will be

$$y(t) \le x_M(t), \quad t \in [t_0, \tilde{t}), \tag{126}$$

with $[t_0, \tilde{t})$ the interval of definition for the function $y(t)$, usually smaller than $[t_0, T)$.

The definition of $x_M(t)$ is

$$x_M(t) = \sup\{x(t); x(t) \le (Vx)(t), \quad x(t) \in S\}, \tag{127}$$

for each $t \in [t_0, T)$. Since $x(t) \in S$, the supremum in the right-hand side of (127) is always finite $(\le z_2(t))$.

We shall prove now that the sup in (127) is the same if instead of $x(t) \in S$, with $x(t) \le (Vx)(t)$, $t \in [t_0, T)$, we take only solutions of (1), i.e., $z(t) = (Vz)(t)$. Indeed, since $VS \subset S$, we can iterate the inequality $x(t) \le (Vx)(t)$, obtaining

$$x(t) \le (Vx)(t) \le (V^2x)(t) \le \cdots \le (V^mx)(t) \le \cdots .$$

These inequalities show that the sequence $\{(V^mx)(t); m \ge 0\}$ is monotonically nondecreasing on $[t_0, T)$. Hence, it does converge (pointwise) on $[t_0, T)$ to some function $z(t) \in S$. But the set $\{(V^mx)(t); m \ge 0\}$ is by our assumption relatively compact in $C([t_0, T), \mathbb{R})$, which means it has a subsequence which converges uniformly on any compact interval of $[t_0, T)$. It is obvious that the limit of this subsequence is $z(t)$. But if a monotonically nondecreasing sequence has a convergent subsequence, the whole sequence is convergent to the same limit. Therefore,

$$\lim_{m \to \infty} (V^mx)(t) = z(t), \tag{128}$$

uniformly on each compact interval of $[t_0, T)$.

From $(V^mx)(t) = V(V^{m-1}x)(t)$, one obtains as $m \to \infty$, $z(t) = (Vz)(t)$, i.e., equation (1).

But $x(t) \le z(t)$ on $[t_0, T)$, which means that for each function satisfying $x(t) \le (Vx)(t)$ on $[t_0, T)$, $x(t) \in S$, we can find a solution $z(t) \in S$ of equation (1). This property tells us that

$$x_M(t) = \sup\{x(t); x(t) = (Vx)(t)\}, \tag{129}$$

which is an alternate definition for $x_M(t)$.

Generally speaking, the supremum taken with respect to a family of continuous functions is not a continuous function (it is a lower semicontinuous function!). But if the family of functions is compact with respect to uniform convergence on compact intervals, then the supremum is also continuous. Indeed, for arbitrary $t_1, t_2 \in [t_0, T)$ we can write

$$x(t_1) \leq |x(t_1) - x(t_2)| + x(t_2).$$

Taking the supremum with respect to all solutions of (1) in S, one obtains

$$\sup x(t_1) \leq \sup |x(t_1) - x(t_2)| + \sup x_2(t).$$

This immediately leads to (by symmetry)

$$|\sup x(t_1) - \sup x(t_2)| \leq \sup |x(t_1) - x(t_2)|.$$

The above inequality means

$$|x_M(t_1) - x_M(t_2)| \leq \sup |x(t_1) - x(t_2)|. \tag{130}$$

The set of solutions to (1) in $C([t_0, T), \mathbb{R})$ is compact, because it is part of a relatively compact set VS, and $x^{(k)}(t) = (Vx^{(k)})(t)$ leads to the closedness of the set of all solutions to (1). Hence, the set of all solutions to (1) is equicontinuous on each compact interval of $[t_0, T)$. Therefore, for any $\varepsilon > 0$, and any interval $[t_0, \tilde{t}]$, $\tilde{t} > \max(t_1, t_2)$, there exists $\delta(\varepsilon) > 0$ such that

$$\sup |x(t_1) - x(t_2)| < \varepsilon \quad \text{for } |t_1 - t_2| < \delta(\varepsilon),$$

which together with (131) yields

$$|x_M(t_1) - x_M(t_2)| < \varepsilon \quad \text{for } |t_1 - t_2| < \delta(\varepsilon),$$

with $t_1, t_2 \in [t_0, \tilde{t}]$.

The continuity of $x_M(t)$ on $[t_0, T)$ being established, it remains to prove that $x_M(t)$ satisfies (1):

$$x_M(t) = (Vx_M)(t), \quad t \in [t_0, T). \tag{131}$$

So far, we know that $x_M(t) \in S$. If one denotes by $z(t)$ the right-hand side of (131), we have for any solution $x(t)$ of (1), $x(t) = (Vx)(t) \leq (Vx_M)(t) = z(t)$. Since $x(t) \leq z(t)$ for any solution $x(t)$ of (1), there follows $x_M(t) \leq z(t)$. We will show that $z(t) \leq x_M(t)$ on $[t_0, T)$, which will imply (131). From the inequality $x_M(t) \leq z(t)$ obtained above, and the monotonicity of V, we have $z(t) = (Vx_M)(t) \leq (Vz)(t)$. Hence, $z(t) \leq (Vz)(t)$ on $[t_0, T)$, which means that $z(t)$ belongs to the class of functions with respect to which one takes the supremum to obtain $x_M(t)$ (the first definition of $x_M(t)$). Consequently, $z(t) \leq x_M(t)$ on $[t_0, T)$, which proves (131).

The existence of the maximal solution for (1) is thus established.

Concerning the minimal solution $x_m(t)$ of (1), we do not need to repeat the argument provided for $x_M(t)$. Indeed, let us denote $-x(t) = y(t)$, $(V_1 y)(t) = -(V(-y))(t)$, which leads to the equation $y(t) = (V_1 y)(t)$, with V_1 continuous and compact, as well as monotonically nondecreasing on the subset of $C([t_0, T), \mathbb{R})$ defined by $-z_2(t) \leq y(t) \leq -z_1(t)$, for all $t \in [t_0, T)$. The equation in y has a maximal solution, as seen above, and this maximal solution for $y(t) = (V_1 y)(t)$ gives the minimal solution for (1), namely, $-y_M(t) = x_m(t)$.

The minimal solution of (1) satisfies the first inequality (125), for all $x(t)$ which are solution of (1).

We shall now return to the functional inequality (121), in order to prove (126).

The answer is almost immediate, because of the (first) definition of $x_M(t)$ by the formula (127). Indeed, the supremum leading to $x_M(t)$ is taken with respect to all $y \in S$, such that $y(t) \leq (Vy)(t)$. Therefore, $y(t) \leq x_M(t), t \in [t_0, T)$.

Remark If $y(t)$ in (121) is defined on $[t_0, \tilde{t}) \subset [t_0, T)$, then (126) is true only on the interval $[t_0, \tilde{t})$.

We can now summarize the result of the discussion conducted above as

Theorem 3.15 *Consider equation (1) and inequality (121), under conditions 1 and 2 formulated above for the operator V. Moreover, assume there exist supersolution and subsolution $z_1(t)$ and $z_2(t)$, satisfying (123) on $[t_0, T)$.*

Then, there exist extremal solutions of (1), $x_m(t)$ and $x_M(t)$, defined on $[t_0, T)$ and satisfying (125) for each solution $x(t)$ of (1). If $y(t) \in C([t_0, T), \mathbb{R})$ satisfies the inequality (121), then (126) is true.

Remark 1 One could assume that $y(t)$ in (121) is defined only on an interval $[t_0, \tilde{t}) \subset [t_0, T)$. Then, (127) is valid only on the interval $[t_0, \tilde{t})$.

Remark 2 A similar statement to that regarding the functional inequality (121) can be obtained for the inequality

$$z(t) \geq (Vz)(t), \quad t \in [t_0, T), \tag{132}$$

with $z(t) \in C([t_0, T), \mathbb{R})$. Then, instead of (126), the following lower estimate is valid:

$$z(t) \geq x_m(t), \quad t \in [t_0, T), \tag{133}$$

where $x_m(t)$ is the minimal solution of (1).

Remark 3 Theorem 3.15 is a tool which allows us to obtain estimates for supersolutions and subsolutions of classical functional (mostly integral) inequalities, given by the extremal solutions of the associate functional equations.

We shall now consider some examples.

The Gronwall integral inequality has the form

$$u(t) \leq f(t) + \int_{t_0}^{t} k(s)u(s)\,ds, \quad t \in [t_0, T), \tag{134}$$

where $f(t)$ and $k(t) \geq 0$ are continuous functions on $[t_0, T)$. The associated equation is the linear integral equation

$$v(t) = f(t) + \int_{t_0}^{t} k(s)v(s)\,ds, \tag{135}$$

for which it is possible to construct a supersolution and subsolution. Let us notice that the right-hand side is monotonically nondecreasing in v, as required in Theorem 3.15. A supersolution of the form

$$\bar{v}(t) = K \exp\left\{\alpha \int_{t_0}^{t} k(s)\, ds\right\}, \quad K, \alpha > 0$$

can be obtained by choosing α and K sufficiently large. Similarly, one can obtain a subsolution to (135).

The equation (135) can be easily reduced to an ordinary differential equation by means of the substitution

$$z(t) = \int_{t_0}^{t} k(s)v(s)\, ds. \tag{136}$$

One obtains $\dot{v}(t) = k(t)[f(t) + v(t)]$, $v(t_0) = 0$. The unique solution of this equation is given by the formula

$$v(t) = \int_{t_0}^{t} f(s)k(s) \exp\left\{\int_{s}^{t} k(u)\, du\right\} ds,$$

which leads to the following estimate for the function $u(t)$ of (134):

$$u(t) \leq f(t) + \int_{t_0}^{t} f(s)k(s) \exp\left\{\int_{s}^{t} k(u)\, du\right\} ds. \tag{137}$$

When $f(t) = M = \text{const.}$ (137) takes a simpler form, namely $u(t) \leq M \exp\{\int_{t_0}^{t} k(s)\, ds\}$.

If (134) is considered on an interval like $[t_0, T)$, $T \leq \infty$, the same estimate (137) is valid because it holds on any compact interval of $[t_0, T)$. Let us notice that in stability problems we will use such estimates on the semi-axis $t \geq t_0$ ($T = +\infty$).

Bihari's integral inequality has the form

$$u(t) \leq M + \int_{t_0}^{t} k(s)g(u(s))\, ds, \quad t \geq t_0, \tag{138}$$

with $M = \text{const.}$ and $k(t) \geq 0$ continuous. The function $g(u)$, in general nonlinear, is assumed continuous and nonnegative, such that $g(u) > 0$ on some semi-axis $u \geq u_0$.

Introducing the function $G(u)$ as a primitive of $1/g(u)$, i.e., $G'(u) = 1/(g(u))$ for $u \geq u_0$, and reducing again (138) to an ordinary differential equation (this time one obtains an equation with separable variables), one obtains the following estimate for any $u(t)$ which satisfies (138):

$$u(t) \leq G^{-1}\left(G(M) + \int_{t_0}^{t} k(s)\, ds\right). \tag{139}$$

G^{-1} is the inverse function of G (i.e., $G(G^{-1}(u)) = u$).

The right-hand side of (139) may not be defined on the interval on which $k(t)$ is assigned. But it is always defined in a neighborhood of t_0, say $t_0 \leq t \leq t_0 + \delta$, $\delta > 0$. It is certainly defined for those t, such that $G(M) + \int_{t_0}^{t} k(s)\, ds$ takes its values in the range of $G(u)$, which is the domain of definition for $G^{-1}(u)$.

In the general case of a Volterra integral equation of the form (6)

$$x(t) = f(t) + \int_{t_0}^{t} K(t, s, x(s)) \, ds, \quad t \in [t_0, T),$$

the key condition for the existence of extremal solutions $x_m(t)$ and $x_M(t)$ is the monotonicity (nondecreasing) of K with respect to the third argument. We have in mind the scalar case, i.e., when x, f and K take their values in $\mathbb{R}$. The result can be extended when x, f and K take values in $\mathbb{R}^n$, considering the usual partial ordering of $\mathbb{R}^n$ ($x \leq y$ iff $x_i \leq y_i, i = 1, 2, \ldots, n$).

The inequality associated with (6) is

$$y(t) \leq f(t) + \int_{t_0}^{t} K(t, s, x(s)) \, ds, \tag{140}$$

and the estimate one finds for $y(t)$ is $y(t) \leq x_M(t), t \geq t_0$, as long as both $y(t)$ and $x_M(t)$ are defined.

Of course, both Gronwall's and Bihari's inequalities are special cases of (140). The reason we have discussed them in detail is due to the fact that, in these cases, we have been able to find the maximal solution $x_M(t)$ and render very effective the use of these inequalities.

Let us point out that Theorem 3.15 allows us to deal with functional differential inequalities of the form

$$\dot{y}(t) \leq (Vy)(t), \quad t \in [t_0, T), \tag{141}$$

in which V is monotonically nondecreasing. From (141) one obtains by integration of both sides, assuming also $y(t_0) \leq y_0$,

$$y(t) \leq y_0 + \int_{t_0}^{t} (Vy)(s) \, ds, \quad t \in [t_0, T), \tag{142}$$

which is a functional inequality of the type discussed in Theorem 3.15. Indeed, the right-hand side of (142) is monotonically nondecreasing in $y(t) \in C([t_0, T), \mathbb{R})$.

Since the maximal solution of the associated equation to (142) also satisfies $\dot{x}_M(t) = (Vx_M)(t)$, we can speak about extremal solutions to the functional differential equation $\dot{x}(t) = (Vx)(t)$, under the monotonicity assumption on V. It is known that this hypotheses is not necessary in the case of ordinary differential equations $\dot{x}(t) = f(t, x(t))$, and the corresponding differential inequalities.

In concluding this section, we will mention that results similar to that in Theorem 3.15 can be obtained for other types of functional or functional differential equations. The reason for dealing with such inequalities is that they provide estimates for the solutions of various types of equations, and are applicable to basic problems such as existence, uniqueness, continuous dependence of solutions with respect to data, stability, etc. Illustrations will be given in subsequent chapters.

3.7 Initial value problems of the second kind for functional differential equations

So far, the initial value problems investigated for functional differential equations of the form (2), $\dot{x}(t) = (Vx)(t)$, have been of the classical (Cauchy) type: $x(t_0) = x^0 \in \mathbb{R}^n$. This was possible because of the causal property of the operator V. And we have seen that other types of initial data, including functional data for delay equations, can be reduced to classical type, adequately defining or redefining the operator V.

It turns out that more general types of initial value problems can be formulated for functional differential equations of the form (2), and they make proper sense and, perhaps, are more interesting from the applications point of view. From a mathematical point of view, such generalizations can be reduced to the classical case, i.e., when we have only point data instead of point and functional data.

The second kind of initial value problem we shall consider now consists in looking for a solution to (2), such that the following conditions are satisfied:

$$x(t) = \varphi(t), \quad t \in [t_0, \tau), \qquad x(\tau) = x^0 \in \mathbb{R}^n, \tag{143}$$

where $t_0 < \tau < T$, and φ is given on $[t_0, \tau)$.

Of course, one can choose $x^0 = \varphi(\tau)$, when φ is continuous on $[t_0, \tau]$. Otherwise, when dealing with measurable functions, for instance, there is no way of matching $\varphi(\tau)$ and x^0, since φ is defined only almost everywhere on $[t_0, \tau]$.

There are at least two reasonable interpretations for the conditions of the form (143): first, determining (by observation) $\varphi(t)$ on $[t_0, \tau)$, and then pursuing the evolution of the process described by equation (2) for $t > \tau$.; second, looking at $\varphi(t)$ as a control function, and choosing it in such a way to achieve certain behavior for the system when $t > \tau$.

To summarize the above discussion about the second kind of initial value problem, we look for a solution of the problem (2), (143), such that $x = x(t; \tau, x^0, \varphi)$ satisfies (2) for $t > \tau$ (possibly only a.e.), $x(t; \tau, x^0, \varphi) = \varphi(t)$ on $[t_0, \tau)$, and $x(\tau; \tau, x^0, \varphi) = x^0 \in \mathbb{R}^n$.

We shall now concentrate on the existence, or existence and uniqueness, of the solution of the problem (2), (143), under adequate conditions. Also, we shall be concerned with properties of the function $x(t; \varphi, x^0)$, assuming τ fixed. Of course, dependence upon τ could also be considered.

We shall choose the space $L^2_{\text{loc}}([t_0, T), \mathbb{R}^n)$ as the underlying space, which means that $V : L^2_{\text{loc}} \to L^2_{\text{loc}}$, and accordingly $\varphi \in L^2([t_0, \tau), \mathbb{R}^n)$.

The following hypotheses will be made on the operator V in the functional differential equation(2), and initial data φ in order to secure the existence of a unique solution satisfying the initial conditions (143):

1   $V : L^2_{\text{loc}}([t_0, T), \mathbb{R}^n) \to L^2_{\text{loc}}([t_0, T), \mathbb{R}^n)$ satisfies a generalized Lipschitz condition of the form

$$\int_{t_0}^t |(Vx)(s) - (Vy)(s)|^2 \, ds \le L(t) \int_{t_0}^t |x(s) - y(s)|^2 \, ds, \tag{144}$$

for all $t \in (t_0, T)$, where $L(t)$ is (necessarily) a nondecreasing nonnegative function on $[t_0, T)$.

2   $\varphi(t) \in L^2([t_0, \tau], \mathbb{R}^n)$. $\tag{145}$

Let us notice that condition (144) implies both causality and continuity of the operator V on the space $L^2_{\text{loc}}([t_0, T), \mathbb{R}^n)$.

Since $x(t; \varphi, x^0)$ must satisfy equation (2) for $t \in [\tau, T)$ only, it is desirable to "shift" the initial time in our considerations from t_0 to τ. Then, conveniently defining an auxiliary operator, associated with V and φ, we will actually investigate existence (and uniqueness in this case) taking into account only the initial condition $x(\tau) = x^0$.

Indeed, to the couple V, φ we shall associate an operator V_φ, defined as follows. For each $x(t) \in L^2_{\text{loc}}([\tau, T), \mathbb{R}^n)$ we define

$$x_\varphi(t) = \begin{cases} \varphi(t), & t \in [t_0, \tau), \\ x(t), & t \in [\tau, T). \end{cases} \tag{146}$$

It is obvious that $x_\varphi(t) \in L^2_{\text{loc}}([t_0, T), \mathbb{R}^n)$. Then, $V_\varphi : L^2_{\text{loc}}([t_0, T), \mathbb{R}^n) \rightarrow L^2_{\text{loc}}([t_0, T), \mathbb{R}^n)$ is defined by

$$(V_\varphi x)(t) = (V x_\varphi)(t). \tag{147}$$

The operator V_φ absorbs the functional initial condition $x(t) = \varphi(t)$ on $[t_0, \tau)$.

The problem that must be solved now, in order to provide an answer to the question of existence (and uniqueness) for equation (2), under conditions (143), can be formulated as follows.

Find the solution of the functional differential equation

$$\dot{x}(t) = (V_\varphi x)(t), \quad t \in [\tau, T), \tag{148}$$

satisfying the initial condition

$$x(\tau) = x^0 \in \mathbb{R}^n. \tag{149}$$

This initial value problem has been investigated earlier in this chapter, under various assumptions. This time, we shall rely on condition 1 above for the operator V, and proceed by successive approximations.

First, let us notice that the operator V_φ in (148) satisfies a Lipschitz type condition similar to (144), namely

$$\int_\tau^t \left| (V_\varphi x)(s) - (V_\varphi y)(s) \right|^2 ds \leq L(t) \int_\tau^t |x(s) - y(s)|^2 ds, \tag{150}$$

for $t \in [\tau, T)$ and any $x, y \in L^2_{\text{loc}}([\tau, T), \mathbb{R}^n)$. The condition (150) follows directly from (144) since both integrands in (144) are zero for $t \in [t_0, \tau)$.

Theorem 3.16 *Consider the problem* (2), (143), *and assume that conditions* 1 *and* 2 *formulated above are satisfied by* V *and* φ.

Then there exists a unique solution $x(t; \tau, x^0, \varphi)$, *locally absolutely continuous on* $[\tau, T)$, *which satisfies* (2) *a.e. for* $t \in [\tau, T)$.

Proof Since (148), (149) are equivalent to the single functional equation

$$x(t) = x^0 + \int_\tau^t (V_\varphi x)(s)\, ds, \quad t \in [\tau, T), \tag{151}$$

we will start the process of successive approximations

$$x^{(m+1)}(t) = x^0 + \int_\tau^t (V_\varphi x^{(m)})(s)\, ds, \tag{152}$$

by letting $x^{(0)}(t) = x^0$. Continuing according to (152), we obtain the sequence $\{x^{(m)}(t); m \geq 0\}$ of locally absolutely continuous functions on $[\tau, T)$.

The following recurrence inequality is a consequence of (152) and (150),

$$|x^{(m+1)}(t) - x^{(m)}(t)|^2 \leq t L^2(t) \int_\tau^t |x^{(m)}(s) - x^{m-1}(s)|^2 \, ds,$$

which is valid for $m \geq 1$ and $t \in [\tau, T)$. If one considers this inequality on a fixed interval $[\tau, \bar{t}], \bar{t} < T$, arbitrary and denote $\bar{t} L^2(\bar{t}) = K$, then we can write

$$|x^{(m+1)}(t) - x^{(m)}(t)| \leq K \int_\tau^t |x^{(m)}(s) - x^{(m-1)}(s)|^2 \, ds$$

for $m \geq 1$ and $t \in [\tau, \bar{t}]$. The above inequality can be processed routinely, leading to the estimate

$$|x^{(m+1)}(t) - x^{(m)}(t)| \leq A K^m \frac{(t - \tau)^m}{m!}, \tag{153}$$

for $t \in [\tau, \bar{t}]$, with $A = \sup |(Vx^0)(t)|$.

The estimate (153) proves the absolute and uniform convergence of the series

$$\sum_1^\infty |x^{(m)}(t) - x^{(m-1)}(t)|$$

on $[\tau, \bar{t}]$, which means that $x^{(m)}(t) \to x(t)$ as $m \to \infty$ uniformly. Therefore, letting $m \to \infty$ in (152), one obtains (151). The uniqueness of the solution follows by the same method, estimating $|y(t) - x^{(m)}(t)|$, with $y(t)$ a solution to (152).

Consequently, Theorem 3.16 is proven.

Remark 1   If instead of (144) one assumes

$$|(Vx)(t) - (Vy)(t)| \leq \lambda(t) \left\{ \int_{t_0}^t |x(s) - y(s)|^2 \, ds \right\}^{1/2}, \tag{154}$$

then the conclusion of Theorem 3.16 remains valid if $\lambda(t) \in L^2_{\text{loc}}$.

Indeed, one derives from the above inequality

$$\int_{t_0}^t |(Vx)(s) - (Vy)(s)|^2 \, ds \leq \int_{t_0}^t \lambda^2(s) ds \int_{t_0}^s |x(u) - y(u)|^2 du$$

$$\leq \left(\int_{t_0}^t \lambda^2(s) \, ds \right) \left(\int_{t_0}^t |x(u) - y(u)|^2 du \right).$$

Therefore, (145) is satisfied with $L(t) = \{\int_{t_0}^t \lambda^2(s) ds\}^{1/2}$. In other words, condition (144) is more general than (154).

Remark 2   The result of Theorem 3.16 can be extended without difficulty to the case of L^p-spaces, $1 < p < \infty$. Only minor changes must be made. For instance, instead of (144) one has to impose the condition

$$\int_{t_0}^{t} |(Vx)(s) - (Vy)(s)|^p\, ds \leq L(t)\int_{t_0}^{t} |x(s) - y(s)|^p\, ds.$$

Instead of the Schwarz inequality, one has to use Hölder's inequality in obtaining the estimates.

Remark 3   The case of the space $C([t_0, T), \mathbb{R}^n)$ as the underlying space has been dealt with in the preceding section. See Theorem 3.13 for existence and uniqueness. The result can be easily adapted to the problem (2), (143).

Let us consider an application of Theorem 3.16 to the case of the integrodifferential equation (32), i.e.,

$$\dot{x}(t) = f\left(t, x(t), \int_{t_0}^{t} k(t, s)x(s)\, ds\right), \quad t \in [t_0, T),$$

under initial conditions (143). In other words, in equation (2) we choose the operator

$$(Vx)(t) = f\left(t, x(t), \int_{t_0}^{t} k(t, s)x(s)\, ds\right), \quad t \in [t_0, T),$$

and we want to find adequate conditions for f, such that V satisfies condition 1 formulated above in this section.

We shall assume that $f : [t_0, T) \times \mathbb{R}^n \times \mathbb{R}^n \to \mathbb{R}^n$ satisfies the Carathéodory conditions, and also condition (109),

$$|f(t, x, y) - f(t, \bar{x}, \bar{y})| \leq \lambda(t)[|x - \bar{x}| + |y - \bar{y}|],$$

with $\lambda(t)$ nonnegative and such that

$$\lambda(t) \in L^{\infty}_{\mathrm{loc}}([t_0, T], \mathbb{R}). \tag{155}$$

Since we want to assure $Vx \in L^2_{\mathrm{loc}}$ for $x \in L^2_{\mathrm{loc}}$, we will make the following assumptions on f and k, besides (109) and (155):

a $f(t, \theta, \theta) \in L^2_{\mathrm{loc}}([t_0, T), \mathbb{R}^n)$.
b $k(t, s)$ is measurable in $\Delta = \{(t, s); t_0 \leq s \leq t < T\}$ and satisfies the square integrability condition

$$\int_{t_0}^{\bar{t}} dt \int_{t_0}^{t} |k(t, s)|^2\, ds < +\infty,$$

for each $\bar{t} < T$. This condition can also be written in the form

$$\int_{t_0}^{t} ds \int_{t_0}^{s} |k(s, u)|^2 du \in L^{\infty}_{\mathrm{loc}}([t_0, T), \mathbb{R}). \tag{156}$$

From the conditions mentioned above, one can see that for each $x(t) \in L^2_{loc}([t_0, T), \mathbb{R}^n)$ one has

$$f\left(t, x(t), \int_{t_0}^t k(t, s)x(s)\, ds\right) \in L^2_{loc}([t_0, T), \mathbb{R}^n).$$

The application of the Krasnoselkii condition for a Niemytskii operator to act continuously on L^2 is satisfied on each $[t_0, \bar{t}] \in [t_0, T)$.

In conclusion, $V = f$ is an operator acting continuously on $L^2_{loc}([t_0, T), \mathbb{R}^n)$, and on each interval $[t_0, \bar{t}] \subset [t_0, T)$ we have

$$\int_{t_0}^t |(Vx)(s) - (Vy)(s)|^2\, ds \le L(\bar{t}) \int_{t_0}^t |x(s) - y(s)|^2\, ds, \tag{157}$$

with $L(t) \in L^\infty_{loc}([t_0, T), \mathbb{R}^n)$.

But (157) is exactly the condition required for the operator V in Theorem 3.16, which means that the integrodifferential equation (32) has a unique solution on $[\tau, T)$, locally absolutely continuous, such that the initial conditions (143) are satisfied.

Remark   The reduction of equation (32), with initial data (143), to the simplest problem when only a point datum is associated, can be done as follows:

$$f\left(t, x(t), \int_{t_0}^t k(t, s)x(s)\, ds\right) = f\left(t, x(t), \int_{t_0}^\tau k(t, s)\varphi(s)\, ds + \int_\tau^t k(t, s)x(s)\, ds\right)$$

$$= \bar{f}\left(t, x(t), \int_\tau^t k(t, s)x(s)\, ds\right).$$

Then, to equation (32), with $\bar{f}$ instead of f, one has to associate only $x(\tau) = x^0 \in \mathbb{R}^n$.

In concluding this section, we shall investigate the problem of continuous dependence of the (unique) solution, with respect to the initial data. Integral inequalities will help us in this respect.

As above, let us denote by $x(t; \tau, x^0, \varphi)$ the solution of (2), satisfying the conditions (143). We know that $x(t; \tau, x^0, \varphi)$ is locally absolutely continuous on $[\tau, T)$, satisfies (2) a.e., but we do not know how x depends on the arguments τ, x^0, φ. Keeping τ fixed, we shall see what kind of dependence we obtain for $x(t; \tau, x^0, \varphi)$, when the operator V satisfies condition 1, formulated above in this section.

Besides the solution $x(t; \tau, x^0, \varphi)$, let us consider another solution of the problem (2), (143). Let us denote it by $y(t; \tau, y^0, \psi)$, which means that the data in (143) are replaced by y^0 and ψ, with $y^0 \in \mathbb{R}^n$ and $\psi \in L^2([t_0, \tau], \mathbb{R}^n)$. We obtain from (2), (143), for x, resp. y, the following equation (after integration):

$$x(t; \tau, x^0, \varphi) - y(t; \tau, y^0, \psi)$$

$$= x^0 - y^0 + \int_\tau^t [(V_\varphi x)(s) - (V_\psi y)(s)]\, ds$$

$$= x^0 - y^0 + \int_\tau^t [(V_\varphi x)(s) - (V_\varphi y)(s)]\, ds + \int_\tau^t [(V_\varphi y)(s) - (V_\psi y)(s)]\, ds$$

$$= x^0 - y^0 + \int_\tau^t [(Vx_\varphi)(s) - (Vy_\varphi)(s)]\, ds + \int_\tau^t [(Vy_\varphi)(s) - (Vy_\psi)(s)]\, ds.$$

We had also to take into consideration (146) and (147). Writting now just $x(t)$ and $y(t)$ instead of $x(t; \tau, x^0, \varphi)$ and $y(t; \tau, y, \psi)$, we obtain from the above equalities by applying also the Schwarz inequality:

$$|x(t) - y(t)| \leq |x^0 - y^0| + (t - \tau)^{1/2} \left\{ \int_\tau^t \left|(V x_\varphi)(s) - (V y_\varphi)(s)\right|^2 ds \right\}^{1/2}$$

$$+ (t - \tau)^{1/2} \left\{ \int_\tau^t \left|(V y_\varphi)(s) - (V y_\psi)(s)\right|^2 ds \right\}^{1/2}.$$

Obviously, we can write

$$|x(t) - y(t)| \leq |x^0 - y^0| + (t - \tau)^{1/2} \left\{ \int_{t_0}^t \left|(V x_\varphi)(s) - (V y_\varphi)(s)\right|^2 ds \right\}^{1/2}$$

$$+ (t - \tau)^{1/2} \left\{ \int_{t_0}^t \left|(V y_\varphi)(s) - (V y_\psi)(s)\right|^2 ds \right\}^{1/2}.$$

This leads, using condition 1 above and (146), (147), to the inequality:

$$|x(t) - y(t)| \leq |x^0 - y^0| + (t - \tau)^{1/2} \left\{ L(t) \int_{t_0}^t \left|x_\varphi(s) - y_\varphi(s)\right|^2 ds \right\}^{1/2}$$

$$+ (t - \tau)^{1/2} \left\{ L(t) \int_{t_0}^t \left|y_\varphi(s) - y_\psi(s)\right|^2 ds \right\}^{1/2}.$$

But

$$\int_{t_0}^t \left|x_\varphi(s) - y_\varphi(s)\right|^2 ds = \int_\tau^t |x(s) - y(s)|^2 ds,$$

and

$$\int_{t_0}^t \left|y_\varphi(s) - y_\psi(s)\right|^2 ds = \int_{t_0}^\tau |\varphi(s) - \psi(s)|^2 ds.$$

Therefore, we can rewrite the above inequality in the form

$$|x(t) - y(t)| \leq |x^0 - y^0| + [(t - \tau)L(t)]^{1/2} \left(\int_\tau^t |x(s) - y(s)|^2 ds \right)^{1/2}$$

$$+ [(t - \tau)L(t)]^{1/2} \left(\int_{t_0}^\tau |\varphi(s) - \psi(s)|^2 ds \right)^{1/2}.$$

If we now rely on the elementary inequality $(a + b + c)^2 \leq 3(a^2 + b^2 + c^2)$, then we can write from the previous inequality for $|x(t) - y(t)|$

$$|x(t) - y(t)|^2 \leq 3 \left| x^0 - y^0 \right|^2 + 3(t - \tau)L(t) \int_\tau^t |x(s) - y(s)|^2 ds$$

$$+ 3(t - \tau)L(t) \int_{t_0}^\tau |\varphi(s) - \psi(s)|^2 ds.$$

One more step will bring the above inequality to the classical Gronwall's form

$$|x(t) - y(t)|^2 \le 3|x^0 - y^0|^2 + 3(t - \tau)L(t) \int_{t_0}^{\tau} |x(s) - y(s)|^2 \, ds$$

$$+ 3(t - \tau)L(t) \int_{\tau}^{t} |x(s) - y(s)|^2 \, ds.$$

This is valid for $t \in [t_0, T)$. If we notice that $(t - \tau)L(t)$ is nondecreasing on $[t_0, T)$, and we fix $\bar{t} < T$, then letting $K = 3(\bar{t} - \tau)L(\bar{t})$ one obtains for $t \in [\tau, \bar{t}]$

$$|x(t) - y(t)|^2 \le 3|x^0 - y^0|^2 + K \int_{t_0}^{\tau} |\varphi(s) - \psi(s)|^2 \, ds + K \int_{\tau}^{t} |x(s) - y(s)|^2 \, ds,$$

which immediately leads to the estimate

$$|x(t) - y(t)|^2 \le \left(3|x^0 - y^0|^2 + K \int_{t_0}^{\tau} |\varphi(s) - \psi(s)|^2 \, ds \right) \exp\{K(t - \tau)\}, \quad t \in [\tau, \bar{t}].$$

The result of this discussion can be summarized in the following theorem.

Theorem 3.17   *Consider the functional differential equation (2), under initial conditions (143). Let $V : L^2_{\text{loc}}([t_0, T), \mathbb{R}^n) \to L^2_{\text{loc}}([t_0, T), \mathbb{R}^n)$ satisfy the condition (144), with $L(t)$ a nonnegative nondecreasing function on $[t_0, T)$, and assume $\varphi \in L^2([t_0, \tau], \mathbb{R}^n)$. The unique solution (see Theorem 3.16) of this equation, with initial conditions (142), say $x(t; \tau, x^0, \varphi)$, is Lipschitz continuous in (x^0, φ), uniformly with respect to t in any compact interval of $[\tau, T)$.*

More precisely, for every $\bar{t} < T$, there exists positive M and K, such that for $t \in [\tau, \bar{t}]$

$$|x(t; \tau, x^0, \varphi) - x(t; \tau, y^0, \psi)|$$

$$\le M \left(|x^0 - y^0| + \left(\int_{t_0}^{\tau} |\varphi(s) - \psi(s)|^2 \, ds \right)^{1/2} \right) \exp\{K(t - \tau)\}.$$

Remark   The number K has been defined above, while M must be such that

$$\left\{ 3|x^0 - y^0|^2 + K \int_{t_0}^{\tau} |\varphi(s) - \psi(s)|^2 \, ds \right\}^{1/2}$$

$$\le M \left(|x^0 - y^0| + \left(\int_{t_0}^{\tau} |\varphi(s) - \psi(s)|^2 \, ds \right)^{1/2} \right).$$

4 Linear and quasilinear equations with causal operators

4.1 Global existence and uniqueness for linear functional differential equations

We shall be concerned first with functional differential equations of the form

$$\dot{x}(t) = (Lx)(t) + f(t), \quad t \in [0, T], \tag{1}$$

with L a linear, causal and continuous operator on a certain function space $E([0, T], \mathbb{R}^n)$. It is possible to build up an existence (and uniqueness) theory, when $L : E \to F$, with F another function space: $F = F([0, T], \mathbb{R}^n)$. In this chapter we agree to take $t_0 = 0$, as initial moment.

For instance, Theorem 3.12 provides an answer to the existence problem, under condition (3.95). But (3.95) is the continuity condition for the operator V (in our case, L) on the space $C([0, T], \mathbb{R}^n)$, assuring also the causality. Therefore, if $L : C([0, T], \mathbb{R}^n) \to C([0, T], \mathbb{R}^n)$ is such that

$$|(Lx)(t) - (Ly)(t)| \leq \lambda(t) \sup_{0 \leq s \leq t} |x(s) - y(s)|, \tag{2}$$

with $\lambda(t)$ a positive nondecreasing function on $[0, T)$, Theorem 3.12 applies and yields the existence and uniqueness for (1), with initial condition $x(0) = x^0$.

Since Lx is a linear operator, (2) is equivalent to

$$|(Lx)(t)| \leq \lambda(t) \sup_{0 \leq s \leq t} |x(s)|, \quad t \in [t_0, T), \tag{2'}$$

which is, as mentioned, the boundedness or continuity condition for L on $C([0, T], \mathbb{R}^n)$.

The same condition of linearity and continuity of the operator L in equation (1), in the spaces $L^p_{\text{loc}}([0, T), \mathbb{R}^n), 1 \leq p < \infty$, assures the existence and uniqueness of a locally absolutely continuous solution on $[0, T)$ to equation (1), with the initial condition $x(0) = x^0 \in \mathbb{R}^n$.

Indeed, we will consider the equivalent problem of finding solutions to the functional integral equation

$$x(t) = x^0 + \int_0^t f(s) \, ds + \int_0^t (Lx)(s) \, ds, \quad t \in [0, T). \tag{3}$$

Of course, besides the continuity of L as an operator from $L^p_{\text{loc}}([0, T), \mathbb{R}^n)$ into itself, we need to assume $f \in L^p_{\text{loc}}([0, T), \mathbb{R}^n)$.

The classical method of successive approximations can be applied, with minor changes. Let us notice that the right-hand side in (3) is locally absolutely continuous for any $x, f \in L^p_{loc}([0, T), \mathbb{R}^n)$.

Consider the sequence defined recurrently by the formula

$$x^{(m+1)}(t) = x^0 + \int_0^t f(s)\,ds + \int_0^t (Lx^{(m)})(s)\,ds, \tag{4}$$

starting, for instance, with $x^{(0)}(t) = x^0$. All terms of this sequence are locally absolutely continuous on $[0, T)$, with values in $\mathbb{R}^n$.

The L^p-continuity condition for L can be written as

$$\int_0^t |(Lx)(s)|^p\,ds \le \lambda(t) \int_0^t |x(s)|^p\,ds, \tag{5}$$

for any $t \in (0, T)$, $x \in L^p_{loc}([0, T), \mathbb{R}^n)$. The function $\lambda(t)$ is positive and nondecreasing on $[0, T)$.

From (4) one obtains

$$|x^{(m+1)}(t) - x^{(m)}(t)|^p \le \left| \int_0^t [(L(x^{(m)} - x^{(m-1)}))(s)]\,ds \right|^p,$$

which shall be discussed separately for $p = 1$ and $1 < p < \infty$. For $p = 1$, taking into account (5), one obtains

$$|x^{(m+1)}(t) - x^{(m)}(t)| \le \lambda(t) \int_0^t |x^{(m)}(s) - x^{(m-1)}(s)|\,ds, \tag{6}$$

for $m \ge 1$ and any $t \in (0, T)$. For $p > 1$, by means of Hölder's inequality, one obtains

$$|x^{(m+1)}(t) - x^{(m)}(t)|^p \le t^{p/q} \lambda(t) \int_0^t |x^{(m)}(s) - x^{(m-1)}(s)|^p\,ds, \tag{7}$$

where $p^{-1} + q^{-1} = 1$.

Both inequalities (6) and (7) can be processed in the same way we have done with inequality (2.87), which leads to the conclusion that the series

$$\sum_{m=0}^{\infty} |x^{(m+1)}(t) - x^{(m)}(t)|$$

is uniformly convergent on any compact interval $[0, \tilde{t}] \subset [0, T)$. If one denotes

$$\lim_{m \to \infty} x^{(m)}(t) = x(t), \tag{8}$$

we find that $x(t)$ satisfies equation (3) on $[0, T)$. Passage to the limit in the integral appearing in (4) is legitimate because uniform convergence on a compact interval implies convergence in any L^p, $1 \le p < \infty$. On the other hand, convergence in L^p, $1 < p < \infty$, on a compact interval, implies convergence in L^1 on the same interval.

The uniqueness of the solution for (1), under L^p conditions specified above and with the usual initial datum $x(0) = x^0$, can be proven also by successive approximations. We have encountered a similar situation in the case of equation (2.84). The recurrent inequality to be processed is

$$|y(t) - x^{(m+1)}(t)|^p \leq \lambda(t) \int_0^t |y(s) - x^{(m)}(s)|^p \, ds,$$

in which $y(t)$ represents a solution of (3), while $x^{(m)}(t)$ are defined by (4).

The discussion carried above, in this section, leads to the following result in regard to the functional differential equation (1), with the initial condition $x(0) = x^0 \in \mathbb{R}^n$.

Theorem 4.1 *Consider the functional differential equation* (1), *in the space* $E([0, T), \mathbb{R}^n)$, *where E stands for any of the spaces* $C([0, T), \mathbb{R}^n)$ *or* $L^p_{loc}([0, T), \mathbb{R}^n)$, $1 \leq p < \infty$.

Assume $f \in E$, *and* $L : E \to E$ *to be linear, causal and continuous (or bounded).*

Then, there exists a unqiue solution of (1), *with initial condition* $x(0) = x^0 \in \mathbb{R}^n$, *defined on* $[0, T)$. *This solution is in the class* $C^{(1)}$ *when* $E = C$, *and it is locally absolutely continuous (hence, satisfies* (1) *a.e.) when* $E = L^p$, $1 \leq p < \infty$.

Remark 1 One can easily see that the solution depends continuously on the data x^0, f. This is done by using integral inequalities (Gronwall) and is an argument of the same nature as in the case of Theorem 3.17.

Remark 2 In Theorem 2.1, the operator L acts on a specific function space (i.e., both the domain and range of L are in the same space). It is possible to discuss a modified version, when the domain is a certain function space, while the range belongs to another function space. Such a situation was encountered in Theorem 3.13.

Remark 3 An existence theorem, similar to Theorem 4.1, can be obtained for the second kind of initial value problem for equation (1). We will discuss this matter in subsequent sections, when dealing with the integral representation of the solutions for linear equations like (1).

4.2 Global existence and uniqueness for linear functional equations

The object of investigation in this section is the functional equation of the form

$$x(t) = (Lx)(t) + f(t), \quad t \in [0, T), \tag{9}$$

with L a linear causal operator on various function spaces, say $E([0, T), \mathbb{R}^n)$. We shall be concerned, as in the preceding section, with the cases $E = C$ or $E = L^p_{loc}$, $1 \leq p < \infty$. Actually, we shall rely on the result given in Theorem 4.1. The method we shall use is that of singular perturbation (see Chapter 3, Section 3).

We associate with (9) the singularly perturbed functional differential equation

$$\varepsilon \dot{x}_\varepsilon(t) = -x_\varepsilon(t) + (Lx_\varepsilon)(t) + f(t), \tag{10}$$

where ε is a small positive parameter. It is necessary to distinguish the continuous case, when L is acting on $C([0, T), \mathbb{R}^n)$ and $f(t)$ is continuous, from the measurable case, when L is acting on L^p_{loc} spaces and $f(t) \in L^p_{loc}$.

From Chapter 2, Section 3, we know that the method of singular perturbations leads to approximate solutions for the unperturbed equation if the operator in the right-hand side is causal, continuous and compact. Therefore, in order to be sure that the solution of (9) is approximated by solutions of (10), it is necessary to assume the compactness of the operator L. Actually, the compactness of L will enable us to show that the family of solutions to (10), say $\{x_\varepsilon(t)\}$, is compact in the underlying space. From this basic property, both the approximation and existence of the solution to (9) will follow by standard arguments.

First, let us notice (see also Chapter 2, Section 3) that equation (10) is equivalent, for each $\varepsilon > 0$, to the functional integral equation

$$x_\varepsilon(t) = c\varepsilon\phi_\varepsilon(t) + \int_0^t \phi_\varepsilon(t-s)[(Lx_\varepsilon)(s) + f(s)]\,ds, \tag{11}$$

where $c \in \mathbb{R}^n$ is an arbitrary vector. As shown by formula (2.66), $\phi_\varepsilon(t) = 0$ for $t < 0$, and $\phi_\varepsilon(t) = \varepsilon^{-1}\exp\{-\varepsilon^{-1}t\}$ for $t \geq 0$. The validity of formula (11) is assured whenever $(Lx_\varepsilon)(t)$ and $f(t)$ belong to the spaces we are interested in: C or L^p, $1 \leq p < \infty$.

Since we are interested in just one solution of (10), we can choose $c = \theta$ in the case of L^p-space, without loss of generality (because $|\exp(\varepsilon^{-1}(t))|_p \to 0$ as $\varepsilon \to 0$). In the case of the space C, $c \in \mathbb{R}^n$ must be chosen as the fixed initial value of the operator $Lx + f$, which is $f(0)$. This can be seen from equation (9), taking into account that $(L\theta)(0) = \theta =$ the fixed initial value for L.

According to the property of the approximate unit with respect to the convolution product, one obtains from (11) and the compactness of the operator L, that $\{x_\varepsilon(t); \varepsilon > 0\}$ is a compact set in L^p-space (resp. the space C). Repeating the argument we have used in Chapter 2, Section 3, we find that there exist sequences $\{x_{\varepsilon_m}(t)\}$, convergent in L^p (or C) to a solution $x(t)$ of equation (9).

It remains now to prove uniqueness of the solution to equation (9), in each space under consideration: C or L^p, $1 \leq p < \infty$. This amounts to the fact that the homogeneous equation associated to (9)

$$y(t) = (Ly)(t), \tag{12}$$

has only the zero solution $x(t) = \theta$. If this is true on any interval $[0, \tilde{t}] \subset [0, T)$, then uniqueness holds on $[0, T)$.

Let us point out that we can restrict our considerations to the case of the space $L^1 = L^1([0, \tilde{t}], \mathbb{R}^n)$, with $\tilde{t} < T$ arbitrarily chosen. Indeed, one has the inclusions $L^p \subset L^1$ for $1 < p < \infty$, as well as $C \subset L^1$.

Hence, we need to show that (12) cannot admit solutions $y(t) \neq \theta$, $y(t) \in L^1([0, \tilde{t}], \mathbb{R}^n)$.

Assume, to the contrary, that there exists a solution of (12) in L^1, which is not identically zero. More precisely, in this case, one can find a smallest number $t_0 \in [0, \tilde{t})$, such that for any $h > 0$, with $t_0 + h \leq \tilde{t}$, one has

$$\int_{t_0}^{t_0+h} |y(s)|\,ds > 0. \tag{13}$$

Otherwise, $y(t)$ should be a.e. equal to θ on $[0, \tilde{t}]$. Without loss of generality, one can assume, for a fixed $h > 0$,

$$\int_{t_0}^{t_0+h} |y(s)|\,ds = 1. \tag{14}$$

Indeed, if $y(t)$ is a solution of (12), then $\lambda y(t)$, with λ real, is also a solution and choosing λ approximately, (13) implies (14).

But $y(t) = \theta$ is a solution of (12) on $[0, \tilde{t}]$, and on each subinterval of $[0, \tilde{t}]$. One can obviously piece together solutions of (12), due to the causality of L, in the following manner.

In case $t_0 > 0$, we will consider a solution of (9) which coincides with θ in $[0, t_0]$. Then, on $[t_0, t_0 + h]$, we will use $y(t)$ satisfying (13) or (14), in order to construct the new solution $\tilde{y}(t)$, such that $\tilde{y}(t) = \theta$ on $[0, t_0]$, and $\tilde{y}(t) = y(t)$ on $[t_0, t_0 + h]$. In case $t_0 = 0$, then $\tilde{y}(t) = y(t)$ on $[0, h]$.

Since L is a compact operator on L^1, for each $\varepsilon > 0$, there exists $\delta > 0$ such that

$$\int_0^{t_0+h} |(Lx)(s-h) - (Lx)(s)|\, ds < \varepsilon = 1, \tag{15}$$

as soon as $h < \delta$, for each $x(t) \in L^1$ satisfying

$$\int_0^{t_0+h} |x(s)|\, ds = 1. \tag{16}$$

In particular, (15) holds true for the solution $\tilde{y}(t)$ considered above, because of (13) and the way we have defined $\tilde{y}(t)$. But $\tilde{y}(t) = (L\tilde{y})(t)$, and therefore (15) implies

$$\int_0^{t_0+h} |\tilde{y}(s-h) - \tilde{y}(s)|\, ds < 1, \tag{17}$$

which reduces in fact to

$$\int_{t_0}^{t_0+h} |\tilde{y}(s)|\, ds = \int_{t_0}^{t_0+h} |y(s)|\, ds < 1. \tag{18}$$

Comparing (14) and (18), we get a contradiction. This proves that our assumption about the existence of a solution $y(t)$ to (9), $y(t)$ not identical to θ, is not acceptable. Therefore, the uniqueness of the solution for equation (9) is true in all function spaces listed above.

In summarizing the discussion carried out in this section, we will state the following theorem.

Theorem 4.2   *Consider equation (9) in the function space $E([0, T), \mathbb{R}^n)$, where E stands for any of the spaces $C([0, T), \mathbb{R}^n)$ or $L_{loc}^p([0, T), \mathbb{R}^n)$, $1 \le p < \infty$.*

Assume $L : E \to E$ is a linear operator, causal, continuous and compact, while $f \in E$. Then, there exists a unique solution of (9) in E.

Remark   It is easy to see that the operator $I - L$, where I stands for the identity operator on E, is invertible. Indeed, equation (9) can be written as $(I - L)x = f$ and it has a unique solution for each $f \in E$. Therefore, $(I-L)^{-1}$ exists on E. On each space $E([0, \tilde{t}], \mathbb{R}^n)$ the inverse operator $(I - L)^{-1}$ is continuous. The solution of (9) can be expressed as $x = (I - L)^{-1}f$.

Let us define the *resolvent* operator associated with L,

$$R = (I - L)^{-1} - I = L(I - L)^{-1}, \tag{19}$$

which leads to the following representation of the solution of (9):

$$x(t) = f(t) + (Rf)(t), \quad t \in [0, T). \tag{20}$$

From (19), it follows that R is also continuous on E. Formula (20) is known as the resolvent formula for (9).

It is useful to notice that the operator R is also a causal operator.

Indeed, if $x(t) = (Lx)(t) + f(t)$ and $y(t) = (Ly)(t) + g(t)$, from $f(t) = g(t)$ on the interval $[0, \tilde{t}]$ we obtain $x(t) - (Lx)(t) = y(t) - (Ly)(t)$ on $[0, \tilde{t}]$, or $x(t) - y(t) = (L(x - y))(t)$ on the same interval. Since uniqueness has been proved for (9), $x(t) = y(t)$ on $(0, \tilde{t}]$. From (20) and the similar formula $y(t) = g(t) + (Rg)(t)$, we obtain $(Rf)(t) = (Rg)(t)$ on $[0, \tilde{t}]$, which proves the causality of the resolvent operator. From (19) it follows that R is also compact.

In Chapter 2, Section 3, Property 6, we have shown how to construct the resolvent operator in the case

$$(Lx)(t) = \int_0^t k(t, s)x(s)\, ds, \tag{21}$$

with $k(t, s)$ continuous on $\Delta = \{(t, s);\ 0 \le s \le t \le T\}$, i.e., when L is the classical Volterra integral operator.

4.3 Integral representation of solutions of linear functional differential equations

Even though the linear equation (1) has a much higher degree of generality than the linear ordinary differential (vector) equation

$$\dot{x}(t) = A(t)x(t) + b(t), \quad t \in [0, T), \tag{22}$$

which we assume is familiar to the reader, it is possible to derive a formula for the solutions of (1), very similar to the classical "variation of parameters" formula. In case of (22), this formula becomes

$$x(t; t_0, x^0) = X(t, t_0)x^0 + \int_{t_0}^t X(t, s)b(s)\, ds, \tag{23}$$

where $X(t, t_0)$ is a matrix of type $n \times n$, whose columns are linearly independent solutions of the homogeneous system $\dot{x}(t) = A(t)x(t)$. In order to estabilish a representation formula, similar to (23), for the solutions of (1) with initial condition $x(0) = x^0 \in \mathbb{R}^n$, we will rely on the continuous dependence of the solution with respect to the data, and the representation of linear continuous functionals on various function spaces. Obviously, this is a matter to be discussed separately, for each specific function space.

We shall start with the case of the space $L^2_{\text{loc}}([0, T), \mathbb{R}^n)$, which appears to be the simplest (perhaps, due to the Hilbert structure on each $L^2([0, \tilde{t}], \mathbb{R}^n)$, $\tilde{t} < T$).

As we already know, equation (1) with initial condition $x(0) = x^0$ is equivalent to equation (3), which we can rewrite as

$$x(t) = g(t) + \int_0^t (Lx)(s)\, ds, \quad t \in [0, T), \tag{24}$$

where

$$g(t) = x^0 + \int_0^t f(s)\, ds, \quad t \in [0, T). \tag{25}$$

The integral term in the right-hand side of (24) can be represented in the form

$$\int_0^t (Lx)(s)\, ds = \int_0^t k(t,s)x(s)\, ds, \tag{26}$$

which follows from the Riesz representation theorem in Hilbert spaces. If one fixes t, $0 < t < T$, then the mapping

$$x(t) \rightarrow \int_0^t (Lx)(s)\, ds$$

is a linear continuous functional on the space $L^2([0,t], \mathbb{R}^n)$, into $\mathbb{R}^n$. Therefore, for each fixed t, the representation (26) is justified, where $k(t,s)$ is square integrable in s on $[0,t]$. Actually, $k(t,s)$ is a matrix kernel of type $n \times n$.

At this point, a theorem of Bukhvalov (see L.V. Kantorovich and G.P. Akilov [1]) is applicable, stating that $k(t,s)$ can be replaced in the representations (26) by a kernel $\tilde{k}(t,s)$ which is measurable in (t,s) on $\Delta = \{(t,s); 0 \le s \le t < T\}$. Therefore, without loss of generality, we can assume that $k(t,s)$ in (26) is a measurable kernel.

Furthermore, the integral operator in the right-hand of (26) takes $L^2([0,\tilde{t}], \mathbb{R}^n)$ into itself (actually, it takes $L^2([0,\tilde{t}], \mathbb{R}^n)$ into the space $AC([0,\tilde{t}], \mathbb{R}^n) \subset L^2$), for each $\tilde{t} < T$. This is possible only if (see, for instance, C. Corduneanu [10, p.37])

$$\int_0^t |k(t,s)|^2\, ds \in C([0,\tilde{t}], R), \quad t \in [0,\tilde{t}],$$

which implies

$$\int_0^{\tilde{t}} dt \int_0^t |k(t,s)|^2\, ds = M^2(\tilde{t}) < \infty, \tag{27}$$

From (24) and (26) it follows that $x(t)$ satisfies the Volterra integral equation

$$x(t) = g(t) + \int_0^t k(t,s)x(s)\, ds, \tag{28}$$

with $g(t)$ given by (25). But condition (7) on the kernel guarantees the existence of a resolvent kernel $\tilde{k}(t,s)$, by means of which the (unique) solution of (28) can be expressed as

$$x(t) = g(t) + \int_0^t \tilde{k}(t,s)g(s)\, ds, \quad t \in [0,T). \tag{29}$$

If we replace $g(t)$ in (29) by its value given by (25), we obtain

$$x(t) = x^0 + \int_0^t f(s)\, ds + \int_0^t \tilde{k}(t,s)\left[x^0 + \int_0^s f(u)\, du\right] ds,$$

which can be rewritten as

$$x(t) = X(t,0)x^0 + \int_0^t X(t,s)f(s)\, ds, \quad t \in [0,\tilde{t}], \tag{30}$$

where

$$X(t, s) = I + \int_s^t \tilde{k}(t, u)\, du. \tag{31}$$

The kernel $\tilde{k}(t, u)$ satisfies a condition similar to (27) and can be constructed as the sum of a series. See F. Tricomi [1] for details.

Let us point out that the first term in the right-hand side of (30) represents the solution of the homogeneous equation $\dot{x}(t) = (Lx)(t)$, with $x(0) = x^0$, while the second term is the solution of (1), such that $x(0) = \theta$.

The matrix $X(t, s)$ is sometimes called the *Cauchy matrix* of the system (1).

Various properties of the matrix $X(t, s)$ can be derived from the properties of $\tilde{k}(t, s)$. We will dwell later on this matter, when such properties will be needed in view of applications of the formula (30).

We will consider one more case of the validity of a formula like (30), considering the situation when the operator L in (1) acts on the space $L^1_{\text{loc}}([0, T), \mathbb{R}^n)$. The discussion of this case is rather similar to the discussion above when L is acting on the space $L^2_{\text{loc}}([0, T), \mathbb{R}^n)$, but different estimates for the quantities involved have to be used.

Assuming now that the operator L in (1) acts continuously on the space $L^1_{\text{loc}}([0, T), \mathbb{R}^n)$, while $f \in L^1_{\text{loc}}([0, T), \mathbb{R}^n)$, we consider again the integral equation (28), in which g is given by (25) and $k(t, s)$ is determined by (26). If necessary, one can again use Bukhvalov's result, and assume without loss of generality that $k(t, s)$ is measurable.

The integral operator in the right-hand side of (26) takes $L^1([0, \tilde{t}], \mathbb{R}^n)$ into the space $C([0, \tilde{t}], \mathbb{R}^n)$ – more precisely, into the space $AC \subset C$. As an operator from $L^1([0, \tilde{t}], \mathbb{R}^n)$ into $C([0, \tilde{t}], \mathbb{R}^n)$, the map $x(t) \to \int_0^t (Lx)(s)\, ds$ is also compact (we have noticed this to be true in Chapter 2, even in the case when L is replaced by a nonlinear operator V which takes bounded sets into bounded sets). According to a result of J. Radon (see, for instance, P. Zabrejko *et al.* [1]), the kernel $k(t, s)$ must satisfy

$$\operatorname*{ess-sup}_{0 \leq s \leq t \leq \tilde{t}} |k(t, s)| = M(\tilde{t}) < \infty, \tag{32}$$

which means, in fact,

$$|k(t, s)| \in L^\infty_{\text{loc}}(\Delta, \mathbb{R}), \tag{33}$$

with $\Delta = \{(t, s); 0 \leq s \leq t < T\}$. We had to take into account the fact that $\tilde{t} < T$ is arbitrarily close to T.

Consequently, the integral equation (28) has its kernel satisfying (33). This feature allows us to construct the resolvent kernel associated with $k(t, s)$, by the same method of series we have used in the case of continuous k. See Chapter 2, Section 6, in which equation (49) with continuous kernel $k(t, s)$ was investigated.

The case under discussion, when instead of continuity in Δ one assumes (33), leads to the construction of the resolvent kernel $\tilde{k}(t, s)$, satisfying a condition similar to (33). The convergence of the series in Chapter 2, Section 6 still holds, but in the L^∞ norm, i.e., ess-sup.

With $\tilde{k}(t, s)$ constructed as above, the (unique) solution of equation (28) is expressed by formula (29). Substituting in (29) the function $g(t)$ given by (25), one obtains again the representation (30) for the solution of (1), with $x(0) = x^0$. Of course, the formula (31) for the Cauchy matrix $X(t, s)$ remains valid under the new assumptions.

Let us notice that this formula can be rewritten as

$$x(t; 0, x^0, f) = X(t, 0)x^0 + \int_0^t X(t, s) f(s) \, ds, \tag{34}$$

while a more general formula, similar to (23), should be valid:

$$x(t; t_0, x^0, f) = X(t, t_0)x^0 + \int_{t_0}^t X(t, s) f(s) \, ds, \tag{35}$$

only for $t \geq t_0 \geq 0$.

Unfortunately this is not true, in general, because $X(t, s)$ may depend on the initial value chosen for t. In the case of (34), the initial value was $t = 0$, while for (35) the initial value is $t = t_0$. If we compare those two formulas, then we see that the value at $t = t_0$, as found from (34), is

$$\bar{x}^0 = X(t_0, 0)x^0 + \int_0^{t_0} X(t_0, s) f(s) \, ds. \tag{36}$$

Constructing the solution for $t > t_0$, taking $\bar{x}^0$ as initial value for the solution, one finds

$$x(t; t_0, \bar{x}^0, f) = X(t, t_0) \left[X(t_0, 0)x^0 + \int_0^{t_0} X(t_0, s) f(s) \, ds \right] + \int_{t_0}^t X(t, s) f(s) \, ds$$

$$= X(t, t_0)X(t_0, t)x^0 + \int_0^{t_0} X(t, t_0)X(t_0, s) f(s) \, ds + \int_{t_0}^t X(t, s) f(s) \, ds.$$

If we would like to equate $x(t; 0, x^0, f)$ with $x(t; t_0, \bar{x}^0, f)$ for $t > t_0$, then we realize that this is possible only when

$$X(t, u)X(u, s) = X(t, s), \tag{37}$$

where $t \geq u \geq s$.

As we know, the property (37) of the Cauchy matrix is valid in the case of ordinary differential equations, when $X(t, s) = X(t)X^{-1}(s)$, with $X(t)$ a fundamental matrix of the homogeneous system associated with (22): $\dot{x}(t) = A(t)x(t)$.

The property expressed by (37) is known as the *semigroup property*. Hence, the semigroup property encountered in the case of ordinary differential equations is not valid, in general, for functional differential equations. This fact certainly represents a drawback, but shows something of the complexity of functional equations.

In the next section, when we discuss the second kind of initial value problem, i.e., the initial problem with functional data (see Chapter 3, Section 7), we will consider a *consistency* problem, which has some relationship with the above-discussed matter.

Let us now see how the formula (30), or equivalently (34), agrees with the classical formula of variation of parameters (Lagrange). As we have seen, this formula is valid under various conditions on the operator L, or the function f, both of which appear in equation (1). We have found that it is valid when L acts on the space L^2, with $f \in L^2$, and also when L acts on L^1, with $f \in L^1$. In Chapter 2, Section 6, we easily found the resolvent operator when $k(t, s)$ is continuous, which means that we also cover the case when we deal with the space C. This is only partially true, because instead of the formula (26), one finds in the case of C a more complex formula:

$$\int_0^t (Lx)(s)\, ds = \int_0^t x(s)\, d_s\mu(t, s), \tag{38}$$

where for each fixed t, $\mu(t, s)$ has bounded variation in s. The above formula follows from the Fréchet theorem of representation of linear continuous functionals on the space $C([0, T], \mathbb{R}^n)$. We shall not treat, in this exposition, the case illustrated by the formula (38). See V.A. Tyshkievich [1] for details.

Returning now to the ordinary differential system (22), we notice that $X(t, s)x^0$ and $X(t, u)X(u, s)x^0$, $s \le u \le t$, are both solutions of $\dot{x}(t) = (Lx)(t)$. Moreover, for $t = s$ both reduce to x^0 (because $X(s, s) = I$), which means that the semigroup property is satisfied, x^0 being arbitrary in $\mathbb{R}^n$. From $X(t, s) = X(t, u)X(u, s), 0 \le s \le u \le t < T$, we obtain for $s = 0$, $X(t, 0) = X(t, u)X(u, 0)$, or $X(t, u) = X(t, 0)X^{-1}(u, 0)$ which explains the well-known factorization of the Cauchy matrix from the theory of linear ordinary differential systems.

Another remarkable application of the general formula (34) is concerned with the functional differential equation with infinite delay. In C. Corduneanu [10], we have treated this particular case in some detail. We shall only briefly discuss this case, without presenting the proof. It does require some technicalities which we do not find necessary to present here.

The equation has the form

$$\dot{x}(t) = L(t, x_t) + f(t), \quad t \in [0, T), \tag{39}$$

where $L(t, x_t)$ stands for a map, linear in the second argument, from $[0, T) \times S$ into $\mathbb{R}^n$, with S the "initial space", i.e., a function space $S = S((-\infty, 0], \mathbb{R}^n)$, satisfying properties described in C. Corduneanu [10]. f is a given function from $[0, T) \to \mathbb{R}^n$, usually locally integrable.

The initial conditions attached to (39) are of the form $x_0(u) = \varphi(u), u \in (-\infty, 0)$, $x_0(0) = x^0$.

The map $L(t, \varphi)$ satisfies a condition of Lipschitz type

$$|L(t, \varphi)| \le \lambda(t)|\varphi|_S, \tag{40}$$

on $[0, T) \times S$, where $\lambda(t)$ is a nonnegative continuous function.

The solution of (39), with assigned initial data, is then represented by

$$x(t) = X(t, 0)x^0 + \int_0^t X(t, s)f(s)\, ds + \int_0^t X(t, s)L(s, \varphi(s+\cdot))\chi_{(-\infty, -s)}(\cdot)\, ds, \tag{41}$$

where $\chi_{(-\infty, -s)}$ is the characteristic function of the semi-axis $(-\infty, -s)$. The formula (41) contains as special cases many formulas of the "variation of parameters" type, including the case of bounded delays. Indeed, if we restrict φ to a subspace of S consisting of functions vanishing for $s \le -h < 0$, then we also cover the case of bounded delays.

4.4 Initial value problems with functional data for linear functional differential equations

We will always refer in this section to the functional differential equation (1), with L a linear causal operator.

As seen in Chapter 3, Section 7, one can associate with (1) the initial conditions

$$x(t) = \varphi(t), \quad t \in [0, \tau), \quad x(\tau) = x^0, \tag{42}$$

and seek the solution for $t > \tau$. This problem is what we have called an initial value problem of the second type.

Since linear functional equations usually provide global existence and uniqueness, we shall denote the solution of (1), with initial conditions (42), by $x(t; \tau, x^0, \varphi)$, as indicated in Chapter 2, Section 7. Sometimes, it is important to emphasize the dependence with respect to f. In such cases, one uses the notation $x(t; \tau, x^0, \varphi, f)$.

In order to derive the existence of the solution to the problem (1), (42), we will rely on Theorem 3.16. As this theorem states, the solution does exist on $[\tau, T)$, as a function of its first argument, is unique, and depends continuously (Lipschitz type continuity) with respect to (x^0, φ). The hypotheses of Theorem 3.16, basically the property 1 of L stated in Chapter 3, Section 7, can be easily verified if we choose the space $L^2_{\text{loc}}([0, T), \mathbb{R}^n)$ as the underlying space, and assume the continuity of the operator L on this space.

Indeed, the continuity condition of L can be written in the form

$$\int_0^t |(L(x - y))(s)|^2 \, ds \le \lambda(t) \int_0^t |x(s) - y(s)|^2 \, ds,$$

which is precisely the condition 1 in Chapter 3, Section 7, with L instead of V. The function $\lambda(t)$ is nonnegative and nondecreasing on $[0, T)$, a feature we have emphasized several times in preceding sections.

Therefore, if the function $\varphi(t) \subset L^2([0, \tau], \mathbb{R}^n)$, all conditions required by Theorem 3.16 are satisfied and the conclusion holds true.

We can say that $x = x(t; \tau, x^0, \varphi)$ is locally absolutely continuous in t on $[0, T)$, and is Lipschitz continuous in (x^0, φ). The dependence on τ has not been investigated in Chapter 3, Section 7, and Theorem 3.17 provides only the properties described above for $x(t; \tau, x^0, \varphi)$.

It is almost obvious that the solution $x(t; \tau, x^0, \varphi, f)$, regarded as a function of the last three arguments, is linear. This means

$$x(t; \tau, x^0, \varphi, f) + x(t; \tau, y^0, \psi, g) = x(t; \tau, x^0 + y^0, \varphi + \psi, f + g). \tag{43}$$

Formula (43) is obtained if we write the equations satisfied by the terms in the left-hand side and then add their first numbers and their second members. Of course, one takes into account the initial data.

One can also see that

$$x(t; \tau, \lambda x^0, \lambda \varphi, \lambda f) = \lambda x(t; \tau, x^0, \varphi, f), \tag{44}$$

for each real λ. One writes the equation for $x(t; \tau, x^0, \varphi, f)$, one multiplies both sides by λ, as well as the initial data, and take into account the uniqueness.

From (43) and (44) we see that the map

$$(x^0, \varphi, f) \to x(t; \tau, x^0, \varphi, f) \tag{45}$$

with t, τ fixed ($0 < \tau < T$) and $t \in [0, \tilde{t}]$, from $\mathbb{R}^n \times L^2([0, \tau], \mathbb{R}^n) \times L^2([0, \tilde{t}], \mathbb{R}^n)$ into $\mathbb{R}^n$, is a linear functional from the direct product of three Hilbert spaces, into $\mathbb{R}^n$. But the product $\mathbb{R}^n \times L^2([0, \tau], \mathbb{R}^n) \times L^2([0, \tilde{t}], \mathbb{R}^n)$ is also a Hilbert space. The continuity of the map (45) has been established only with respect to (x^0, φ), and as Theorem 3.17 says, this is a Lipschitz type continuity. If we prove continuity with respect to the argument f, for fixed x^0 and φ, then the continuity of the map (x^0, φ, f) is proven.

In order to show that the map (45) is continuous in f, we need to consider the equations (similar to (1))

$$\dot{x}(t) = (Lx)(t) + f^{(m)}(t), \qquad (46)$$

with $f^{(m)}(t) \to f(t)$ as $m \to \infty$, in $L^2_{\mathrm{loc}}([0, T), \mathbb{R}^n)$. The initial conditions (42) remain the same for each m. In other words, for (46) we choose the solution $x(t; \tau, x^0, \varphi, f^{(m)})$. We shall prove that

$$x(t; \tau, x^0, \varphi, f^{(m)}) \to x(t; \tau, x^0, \varphi, f), \qquad (47)$$

as $m \to \infty$, uniformly on $[\tau, \tilde{t}]$. If one denotes by $x^{(m)}(t)$ the solution $x(t; \tau, x^0, \varphi, f^{(m)})$, and by $x(t)$ the solution of (1), (42), then

$$x^{(m)}(t) - x(t) = \int_\tau^t \{(L(x^{(m)} - x))(s)\}\,ds + \int_\tau^t \{f^{(m)}(s) - f(s)\}\,ds.$$

This leads immediately to a Gronwall type integral inequality, and observing that

$$\left| \int_\tau^t \{f^{(m)}(s) - f(s)\}\,ds \right|^2 \leq (\tilde{t} - \tau) \int_\tau^t \left| f^{(m)}(s) - f(s) \right|^2 ds,$$

one obtains (47).

Therefore, $x(t; \tau, x^0, \varphi, f)$ is continuous in (x^0, φ, f) on the product space $\mathbb{R}^n \times L^2([0, \tau], \mathbb{R}^n) \times L^2([\tau, \tilde{t}], \mathbb{R}^n)$, which tells us that the map (45) is a linear continuous functional (with values in $\mathbb{R}^n$), for each fixed $t > \varepsilon$, $t < T$.

By the representation theorem of Riesz, for linear continuous functionals on Hilbert space, one obtains

$$x(t; \tau, x^0, \varphi, f) = X(t, \tau)x^0 + \int_0^\tau \tilde{X}(t, s; \tau)\varphi(s)\,ds + \int_\tau^t X(t, s)f(s)\,ds, \qquad (48)$$

which, compared to the formula of "variation of parameters" (34), displays an extra term whose presence indicates the influence of the initial functional datum. For $t = \tau$, the right-hand side of (48) becomes $x^0 + \int_0^\tau \tilde{X}(\tau, s; \tau)\varphi(s)\,ds$. Since this must be x^0, it follows that the second term vanishes for each $\varphi \in L^2$. This means $\tilde{X}(\tau, s; \tau) = \theta$, a.e.

Formula (48) needs some discussion and clarifications, before we can state the basic result about the representation of solutions of the problem (1), (42).

First, we have used the Cauchy matrix $X(t, s)$ in the first and third terms in the right-hand side of (48). This fact is motivated by the reason to reobtain formula (34), with τ as initial moment instead of 0, in case we choose $\varphi = \theta$ (then, the middle term in (48) vanishes). Actually, one can proceed as in case of equation (1), with the point datum $x(\tau) = x^0$, also taking into account the functional datum φ, which brings the middle term in (48).

Second, both $X(t, s)$ and $\tilde{X}(t, s; \tau)$ are square integrable in s. $X(t, s)$ on each interval $[\tau, \tilde{t}]$, for any fixed $t \in [\tau, T)$, and $\tilde{X}(t, s; \tau)$ on $[0, \tau]$. This follows directly from the Riesz representation of linear continuous functional on L^2-spaces.

Third, the formula (48) emphasizes the solution of the homogeneous equation associated to (1), $\dot{x}(t) = (Lx)(t)$, under initial conditions (42). The sum of the first two terms in the right-hand side of (47) is precisely this solution. The last term in the right-hand side of (48) represents the solution of (1), under initial conditions $x(t) = \theta, t \in [0, \tau]$.

Fourth, $X(t, s)$ and $\tilde{X}(t, s; \tau)$ are locally absolutely continuous on $[\tau, T)$ with respect to t, which implies that they are a.e. differentiable.

Fifth, the Cauchy matrix $X(t, s)$ depends, in general, also on τ. We have noticed this feature in the preceding section, when discussing the validity of formula (35).

Summarizing the discussion carried out in this section, we can state the following result in regard to the problem (1), (42).

Theorem 4.3   *Assume that the operator L in (1) is linear, causal and continuous on $L^2_{\text{loc}}([0, T), \mathbb{R}^n)$, while $f \in L^2_{\text{loc}}([0, T), \mathbb{R}^n)$ and $\varphi \in L^2([0, \tau], \mathbb{R}^n)$.*

Then, the solution of the problem (1), (42) can be represented by the formula (48), with $X(t, s)$ and $\tilde{X}(t, s; \tau)$ enjoying the properties described above.

Related to the second kind of initial value problems, we shall discuss now a *consistency* problem which can be formulated as follows.

We consider the equation (1) with initial condition (42), whose solution exists on $[0, T]$, is unique and satisfies the equation a.e. on (τ, T). This solution has been denoted by $x(t; \tau, x^0, \varphi)$.

Let us consider now another initial value problem of the second kind, also for equation (1), but instead of (42) we choose the conditions

$$x(t) = \tilde{\varphi}(t), \quad t \in [0, \tau'), \quad x(\tau') = \tilde{\varphi}(\tau'),$$

where $\tau < \tau' < T$, and

$$\tilde{\varphi}(t) = \begin{cases} \varphi(t), & t \in [0, \tau), \\ x(t; \tau, x^0, \varphi), & t \in [\tau, \tau'). \end{cases}$$

The *consistency* requirement is expressed by the equality $x(t; \tau, x^0, \varphi) = x(t; \tau', \tilde{\varphi}(\tau'), \tilde{\varphi})$, which must be valid for $t > \tau'$.

Of course, the consistency problem has meaning only in case uniqueness holds for the second kind of initial value problem. Otherwise, it does not make sense to have $x(t; \tau, x^0, \varphi) = x(t; \tau', \tilde{\varphi}(\tau'), \tilde{\varphi})$ for $t > \tau'$, because there is a possibility of having distinct solutions of (1) issuing from the point $(\tau', \tilde{\varphi}(\tau'))$.

Since uniqueness holds true for linear equation (1), with L as in Theorem 3.17, we realize that those two solutions must coincide. Otherwise, the initial value problem (1), (42) would have more than one solution. One is $x(t; \tau, x^0, \varphi)$, while the second coincides with $x(t; \tau, x^0, \varphi)$ on $[\tau, \tau']$, and with $x(t; \tau', \tilde{\varphi}(\tau'), \tilde{\varphi}) \neq x(t; \tau, x^0, \varphi)$ for $t \in (\tau', T)$.

Therefore, for linear functional equations the consistency problem has a positive answer.

4.5 Quasilinear functional equations

This section is concerned with quasilinear functional equations or functional differential equations involving causal operators. In the case of functional equations we will consider equations of the form

$$x(t) = (Lx)(t) + (Nx)(t), \quad t \in [0, T], \tag{49}$$

while for functional differential equations we will deal with the analogous equation

$$\dot{x}(t) = (Lx)(t) + (Nx)(t), \quad t \in [0, T]. \tag{50}$$

In both equations (49) and (50), L is a linear causal operator, while N is causal, and in general nonlinear. Of course, we can approach the existence problem in different function spaces. In the case of equation (50), we will associate the initial condition $x(0) = x^0 \in \mathbb{R}^n$. The second kind of initial value problem could also be investigated for the functional differential equation (50).

A feature that will be present in the investigation is the reliance on the properties of the linear "part" of either equation (49) or (50), in order to derive properties of the quasilinear counterpart.

Noteworthy special cases of equation (49) are the perturbed integral equations of the form

$$x(t) = \int_0^t k(t, s) x(s) \, ds + \int_0^t K(t, s, x(s)) \, ds, \tag{51}$$

while for equations of the form (50) one can mention classical integrodifferential equations such as

$$\dot{x}(t) = A(t) x(t) + \int_0^t K(t, s, x(s)) \, ds, \tag{52}$$

with the usual initial condition $x(0) = x^0$.

Equations like (51) or (52) may directly appear in modeling real phenomena from the physical sciences, engineering, economics or society, or may be generated by emphasizing the linear part in such equations.

We shall start with equation (50), with the initial condition $x(0) = x^0 \in \mathbb{R}^n$. The following conditions will be imposed on the operators L and N:

1   L is like in Theorem 4.1;
2   $N : E \to E$ is an operator on the space E, satisfying a Lipschitz type condition

$$|(Nx)(t) - (Ny)(t)| \le \lambda |x(t) - y(t)|, \tag{53}$$

with sufficiently small λ.

Let us notice that condition (53), where the norm is the $\mathbb{R}^n$-norm, implies causality of the operator N. Of course, when E is a space of measurable functions on $[0, T]$, with values in $\mathbb{R}^n$, condition (53) is valid only a.e.

Theorem 4.4   *Under conditions 1 and 2 stated above, the equation (50), with initial condition $x(0) = x^0 \in \mathbb{R}^n$ has a unique solution in the space $E([0, T], \mathbb{R}^n)$. This solution* is of class $C^{(1)}$ when $E = C([0, T], \mathbb{R}^n)$, and is absolutely continuous on $[0, T]$ when $E = L^p([0, T], \mathbb{R}^n)$, $1 \leq p < \infty$.

Proof   Taking the space E as the underlying space, one defines on this space an operator $A : E \to E$, $x(t) = (Au)(t)$, where u and x are related by means of the equation

$$\dot{x}(t) = (Lx)(t) + (Nu)(t), \quad t \in [0, T], \tag{54}$$

with $x(0) = x^0 \in \mathbb{R}^n$. As established in Theorem 4.1, there exists a unique (global) solution of (54), because $Nu \in E$ for $u \in E$. Of course, this solution is absolutely continuous on $[0, T]$, which means that it is in E for each choice of E mentioned above.

Let us consider also $y(t) = (Av)(t)$, for $v \in E$, and denoting $z(t) = x(t) - y(t)$ one obtains from (54) and the similar equation for the couple y, v,

$$z(t) = \int_0^t (Lz)(s) \, ds + \int_0^t [(Nu)(s) - (Nv)(s)] \, ds, \tag{55}$$

taking into account that $x(0) = x^0 = y(0)$.

The operator

$$z(t) \to z(t) - \int_0^t (Lz)(s) \, ds \tag{56}$$

is invertible on E (see Theorem 4.1) and its inverse $R = (I - L)^{-1}$ is, therefore, continuous. Hence, we can rewrite (55) in the form

$$z(t) = R \left(\int_0^t [(Nu)(s) - (Nv)(s)] \, ds \right), \tag{57}$$

which enables us to obtain an estimate for $z(t) = x(t) - y(t) - (Au)(t) - (Av)(t)$. Since both $u, v \in E$, we will need an estimate in this space. Denoting by $|R|_E$ the norm of R on the space E, we obtain from (57) the following inequality:

$$|z|_E \leq |R|_E \left| \int_0^t [(Nu)(s) - (Nv)(s)] \, ds \right|_E. \tag{58}$$

Now using condition (53) for N, we can write

$$|z|_E \leq \lambda |R|_E \left| \int_0^t |u(s) - v(s)| \, ds \right|_E. \tag{59}$$

In case $E = C$, we have for $u, v \in C$

$$\left| \int_0^t |u(s) - v(s)| \, ds \right|_C \leq T|u - v|_C, \tag{60}$$

which combined with (59) leads to

$$|Au - Av|_C \leq \lambda T |R|_C |u - v|_C. \tag{61}$$

Hence, if we choose λ such that $\lambda < (T|R|_C)^{-1}$, the operator A is a contraction and Theorem 4.4 holds true.

In case $E = L^1$, from (59) one derives

$$|Au - Av|_1 \leq \lambda |R|_{L^1} |u - v|_1, \tag{62}$$

which shows that in case $E = L^1$ we need only to choose λ such that $\lambda < (|R|_{L^1})^{-1}$.

If $E = L^p$, $1 < p < \infty$, in the right-hand side of (59) we have to apply the Hölder inequality to the last factor, which leads to

$$|Au - Av|_p \leq \lambda (T)^{1/q} |R|_{L^p} |u - v|_p, \tag{63}$$

with q such that $p^{-1} + q^{-1} = 1$. Again, from (63) we see that A is a contraction on L^p if one chooses $\lambda < (|R|_{L^p})^{-1} (T)^{-1/q}$, which proves Theorem 4.4 in case $E = L^p$, $1 < p < \infty$.

In conclusion, the operator A has a unique fixed point in E, provided λ is sufficiently small. More precisely, one must take λ such that the operator A is a contraction on E. But the fixed point of A is obviously a solution of the functional differential equation (1), satisfying the initial condition $x(0) = x^0 \in \mathbb{R}^n$.

Remark   If instead of the closed interval $[0, T]$ one considers equation (1) on the semi-open interval $[0, T)$, $T \leq \infty$, then the above argument holds on any interval $[0, \tilde{t}] \subset [0, T)$. Moreover, if one compares the solution of (1), with $x(0) = x^0$, on $[0, \tilde{t}]$, with that on $[0, \tilde{t} + h]$, because of the Lipschitz type continuity of the right side of (1) one easily finds that the latter coincides with the first on $[0, \tilde{t}]$. In other words, we can obtain existence on $[0, T)$. One can also proceed by successive approximations, which converge uniformly in the case of the space C, as well as in L^p-spaces, $1 \leq p < \infty$.

Let us consider an application of Theorem 4.4 to equation (52). We will take the space $C([0, T], \mathbb{R}^n)$ as the underlying space. The continuity of $A(t)$ is then a natural assumption. The operator $(Lx)(t) = A(t)x(t)$ is linear, causal and continuous on the space C. We need to make such assumptions on $K(t, s, x)$, in order to assure the integral operator in the right-hand side of (52) is acting on $C([0, T], \mathbb{R}^n)$. Again, one achieves the objective assuming $K(t, s, x)$ is continuous or $\Delta \times \mathbb{R}^n$, $\Delta = \{(t, s); \ 0 \leq s \leq t \leq T\}$, with values in $\mathbb{R}^n$. Then, the integral operator in the right-hand side of (52) acts continuously on $C([0, T], \mathbb{R}^n)$. Actually, it is possible to assume less than continuity on K, but we shall not dwell on this matter here.

Conditions 1 and 2, admitted in Theorem 4.4, can now be easily checked if, besides continuity of $A(t)$ and $K(t, s, x)$, one assumes

$$|K(t, s, x) - K(t, s, y)| \leq \lambda |x - y|, \tag{64}$$

for $(t, s) \in \Delta$ and $x, y \in \mathbb{R}^n$.

Another type of condition to substitute for (64) is

$$\int_0^t |K(t, s, x(s)) - K(t, s, y(s))| \, ds \leq \lambda \sup_{0 \leq s \leq t} |x(s) - y(s)|, \tag{65}$$

for $t \in (0, T]$, and $x, y \in C([0, T], \mathbb{R}^n)$.

Under each of the assumptions (64) or (65), λ will be chosen small enough.

Before we move on to the investigation of the functional equation (49), let us provide another result for equation (50), this time without the continuity assumption on all functions involved. It can be stated as follows.

Theorem 4.5   *Consider equation* (50), *with L a linear, causal and continuous operator on $L^2([0, T], \mathbb{R}^n)$. Assume further that $N : C([0, T], \mathbb{R}^n) \rightarrow L^2([0, T], \mathbb{R}^n)$ is Lipschitz continuous, i.e.,*

$$|Nx - Ny|_2 \leq \lambda |x - y|_C, \tag{66}$$

with λ sufficiently small. Then, there exists a unique solution $x(t) \in AC([0, T], \mathbb{R}^n) \subset C([0, T], \mathbb{R}^n)$ of equation (50), *satisfying the initial condition $x(0) = x^0 \in \mathbb{R}^n$.*

The proof can be carried out by means of the contraction mapping principle, taking the space $C([0, T], \mathbb{R}^n)$ as the underlying space. It is useful to notice that the problem considered above is equivalent to the functional integral equation

$$x(t) = X(t, 0)x^0 + \int_0^t X(t, s)(Nx)(s)\, ds, \tag{67}$$

where $X(t, s)$ is the Cauchy matrix of the linear system $\dot{x}(t) = (Lx)(t)$. The matrix $X(t, s)$ was constructed in Section 3. The linear operator

$$f(t) \rightarrow \int_0^t X(t, s)f(s)\, ds$$

is continuous (and compact) from $L^2([0, T], \mathbb{R}^n)$ into $C([0, T], \mathbb{R}^n)$. This property implies estimates like

$$\int_0^t |X(t, s)|^2\, ds \leq M^2 < \infty, \quad t \in [0, T],$$

or

$$\int_0^t |X(t, s) - X(t_0, s)|^2\, ds < \varepsilon, \quad \text{for } |t - t_0| < \delta(\varepsilon).$$

These are very useful in the application of the contraction mapping principle.

We leave the details of the proof of Theorem 4.5 to the reader.

Let us now return to the functional equation (49), to establish an existence result. This time we will use the Schauder fixed point theorem, which does not provide for the uniqueness of the solution.

Theorem 4.6   *Consider the functional equation* (49), *under the following conditions:*

1   *The operator $L : C([0, T], \mathbb{R}^n) \rightarrow C([0, T], \mathbb{R}^n)$ is linear, causal, continuous and compact (it is understood that $(L\theta)(t) = \theta$ on $[0, T)$).*
2   *The operator $N : C([0, T], \mathbb{R}^n) \rightarrow C([0, T], \mathbb{R}^n)$ is continuous and compact, and defining for $r > 0$*

$$\phi(r) = \sup\{|Nx|_C;\ x \in C,\ |x|_C \leq r\},$$

one has

$$\limsup_{r \to \infty} \frac{\phi(r)}{r} = \lambda,$$

with λ sufficiently small.

Then, there exists a solution $x(t) \in C([0, T], \mathbb{R}^n)$ of equation (49).

Proof   As we have shown in Remark 1 to Theorem 4.2, under our conditions the operator $x(t) \to x(t) - (Lx)(t)$ is invertible and there exists a resolvent operator R on $C([0, T], \mathbb{R}^n)$, also linear, causal, continuous and compact, such that the (unique) solution of the linear equation $x(t) = (Lx)(t) + f(t)$ is represented by

$$x(t) = f(t) + (Rf)(t), \quad t \in [0, T]. \tag{68}$$

From (49) and (68) we obtain the following functional equation for the solution of (49):

$$x(t) = (Nx)(t) + (RNx)(t), \quad t \in [0, T], \tag{69}$$

which is of the form $x(t) = (Vx)(t)$. Let us notice that the right-hand side of (69) defines a continuous and compact operator on the space $C([0, T], \mathbb{R}^n)$. Consequently, the only thing we have to prove in order to use Schauder's fixed point theorem is the existence of a convex closed set in C, which is taken into itself by the operator

$$(Ux)(t) = (Nx)(t) + (RNx)(t).$$

Taking condition 2 of Theorem 4.6 into account, we obtain the following estimate for U, when $|x(t)| \le r$ on $[0, T]$:

$$|(Vx)(t)| \le \phi(r) + |R|_C \phi(r), \quad t \in [0, T]. \tag{70}$$

If we want to assure that the ball B_r, centered at $\theta \in C$, is taken into itself by the operator U, $UB_r \subset B_r$, then we must find $r > 0$, such that

$$(1 + |R|_C)\phi(r) \le r, \tag{71}$$

according to the estimate (70). If we divide both sides of (71) by $r(1 + |R|_C)$ we obtain

$$\frac{\phi(r)}{r} \le (1 + |R|_C)^{-1}. \tag{72}$$

In order to secure the existence of at least one $r > 0$ with property (72), it suffices to assume that

$$\limsup_{r \to \infty} \frac{\phi(r)}{r} = \lambda < (1 + |R|_C)^{-1}. \tag{73}$$

Inequality (73) shows how small one has to choose λ in condition 2 of Theorem 4.6, in order to secure the validity of this theorem.

This ends the proof of Theorem 4.6.

Remark   It appears from the examination of the proof of Theorem 4.6 that the causality of the operator N is not necessary. If we would like to obtain a result based on the causality of U, which certainly holds if N is causal, then we should look at the hypotheses of Theorem 3.7 or Theorem 3.11, and adapt them to equation (2).

In concluding this section, we will apply Theorem 4.6 to the integral equation (51), which appears as a nonlinear perturbation of the linear integral equation of Volterra type

$$x(t) = \int_0^t k(t, s)x(s) + f(t).$$
(74)

We will again take the space $C([0, T], \mathbb{R}^n)$ as the underlying space, and assume that the kernel $k(t, s) = (k_{ij}(t, s))$ of type $n \times n$ is continuous on $\Delta = \{(t, s);\ 0 \le s \le t \le T\}$.

By preserving the notation used in Theorem 3.6, we have

$$(Nx)(t) = \int_0^t K(t, s, x(s))ds,\ t \in [0, T].$$
(75)

If we now assume the continuity of $K(t, s, x)$ in $\Delta \times \mathbb{R}^n$, this will imply the continuity and compactness of N on $C([0, T], \mathbb{R}^n)$. These properties follow easily if we keep in mind that $K(t, s, x)$ is uniformly continuous on any set $\Delta \times B$, with $B \subset \mathbb{R}^n$ a bounded set.

We will further assume that

$$|K(t, s, x)| \le k_0(t, s)m(|x|),$$
(76)

on $\Delta \times \mathbb{R}^n$, where $k_0(t, s)$ is continuous and nonnegative on Δ, while $m(r)$ is positive and nondecreasing on $[0, \infty)$.

Let us denote

$$\gamma = \sup \int_0^t k_0(t, s)ds,\quad 0 \le t \le T.$$
(77)

Then, it is easy to see that

$$\phi(r) \le \gamma m(r),\quad r \in (0, \infty),$$
(78)

and letting

$$\limsup_{r \to \infty} \frac{m(r)}{r} = \rho,$$
(79)

assuming also that $\rho < \infty$, the condition (73) becomes in this case

$$\limsup_{r \to \infty} \frac{\phi(r)}{r} = \lambda \le \gamma \rho < (1 + |R|_C)^{-1}.$$
(80)

The last inequality in (80) will be satisfied if either of the numbers γ and ρ is sufficiently small.

In conclusion, the perturbed integral equation (51), with $k(t, s)$ and $K(t, s, x)$ continuous functions on their domains, has a solution in $C([0, T], \mathbb{R}^n)$ if at least one of the numbers γ and ρ, defined by (76)–(79), is sufficiently small.

Let us point out that the inequality (76) is not a real restriction on $K(t, s, x)$. Indeed, for fixed $(t, s) \in \Delta$, one denotes

$$\sup_{|x| \leq r} |K(t, s, x)| = k_0(t, s)m(r),$$

where both k_0 and m exist. Without loss of generality, one can assume both k_0 and m to be continuous.

4.6 Two-point boundary value problems

Until now, we have been concerned only with initial value problems for functional differential equations. The existence theorems given in Chapter 3, or earlier in this chapter provide conditions assuring the existence of solutions satisfying, besides the functional differential equation $\dot{x}(t) = (Vx)(t)$, the classical initial condition $x(0) = x^0$, or the second kind of initial condition $x(t) = \varphi(t)$, $t \in [0, \tau)$, $x(\tau) = x^0$.

For linear functional equations with causal operators, as well as for quasilinear ones, it is possible to investigate the existence of solutions satisfying a two-point boundary value condition. More precisely, in the linear case we will look for solutions of equation (1), such that

$$Ax(0) + Bx(T) = c, \tag{81}$$

where A and B are square matrices of type $n \times n$, while $c \in \mathbb{R}^n$.

In order to examine the above problem, we will use the integral representation formula for the solutions of equation (1), trying to determine the initial value x^0 by imposing to the solution the boundary value condition (81). Formula (30),

$$x(t) = X(t, 0)x^0 + \int_0^t X(t, s)f(s)\,ds,$$

leads to the following equation for x^0:

$$[A + BX(T, 0)]x^0 = c - B \int_0^T X(T, s)f(s)\,ds. \tag{82}$$

(82) is a linear system for x^0, which will possess a unique solution for any $c \in \mathbb{R}^n$ and $f \in L^2([0, T], \mathbb{R}^n)$ if, and only if,

$$\det[A + BX(T, 0)] \neq 0. \tag{83}$$

If (83) holds true, then we can say that the boundary value problem (1), (81) has a unique solution in $AC([0, T], \mathbb{R}^n)$. We can actually find an integral representation for this unique solution if we substitute x^0 from (82) into the formula (30) for $x(t)$. One obtains the representation

$$x(t) = g(t) + \int_0^T \tilde{X}(t, s)f(s)\,ds, \tag{84}$$

where

$$g(t) = X(t, 0)[A + BX(T, 0)]^{-1}c, \tag{85}$$

and

$$\tilde{X}(t, s) = \begin{cases} X(t, s) - X(t, 0)[A + BX(T, 0)]^{-1}BX(T, s), & 0 \le s \le t, \\ -X(t, 0)[A + BX(T, 0)]^{-1}BX(T, s), & t \le s \le T. \end{cases}$$

We shall summarize the above discussion in the following result.

Theorem 4.7 *Consider the functional differential equation* (1), *in which* $L: L^2([0, T], \mathbb{R}^n) \to L^2([0, T], \mathbb{R}^n)$ *is a linear, causal and continuous operator, and* $f \in L^2([0, T], \mathbb{R}^n)$.

If condition (83) *is satisfied, then there exists a unique solution of* (1), *satisfying the boundary value condition* (81). *This solution belongs to* $AC([0, T], \mathbb{R}^n)$ *and can be represented by formula* (84).

Remark The usual initial value condition $x(0) = x^0$ is a special case of (81). Indeed, it corresponds to $A = I$, $B = 0$, and $c = x^0$. Then, (84) reduces to (30), which show that the classical initial value problem is a special case of (1), (81).

If one chooses $A = I$, $B = -I$ and $c = \theta$, then (81) becomes $x(0) = x(T)$. This is the boundary value condition specific to the periodicity (of the solution). However, we cannot talk about periodic solutions, unless we extend L to act on functions defined on $[0, \infty)$, and f by the periodicity relation $f(t + T) = f(t)$, a.e. on $[0, \infty)$.

The quasilinear counterpart of the boundary value problem (1), (81) can be formulated as follows.

Find solutions of the functional differential equation

$$\dot{x}(t) = (Lx)(t) + (fx)(t), \quad t \in [0, T], \tag{86}$$

satisfying the boundary value condition

$$Ax(0) + B_x(T) = c(x), \tag{87}$$

where L is like in Theorem 4.7, $f : L^2([0, T], \mathbb{R}^n) \to L^2([0, T], \mathbb{R}^n)$ is, in general, a nonlinear operator, continuous and satisfying some extra conditions to be specified below, while $c : C([0, T], \mathbb{R}^n) \to \mathbb{R}^n$ is subject to restrictions we shall describe later.

Both equation (86) and boundary value condition (87) can be termed quasilinear.

This time it is not possible to find x^0, even under assumption (83). Instead, we find a functional equation for x, namely

$$x^0 = [A + BX(T, 0)]^{-1} \left\{ c(x) - B \int_0^T X(T, s)(fx)(s) \, ds \right\}.$$

Substituting this x^0 in formula (30) one finds a functional integral equation for $x(t)$:

$$x(t) = (gx)(t) + \int_0^T \tilde{X}(t, s)(fx)(s) \, ds, \tag{88}$$

and $\tilde{X}(t, s)$ is defined as in equation (84) (i.e., as in the linear case treated in Theorem 4.7). The functional integral equation (88) is of the same nature as equation (69) in the preceding section. One can proceed by the fixed point method, under conditions similar to those encountered above, imposing some groth restricting on $c(x)$ and $f(x)$.

We will not get into details about equation (88) here. We notice only that a growth condition of the form (80) for f, and a Lipschitz type condition for $c(x)$, with sufficiently small constant, will assure the existence of a solution to (88), i.e., to the boundary value problem (86), (87).

4.7 Comments and references

The theory of linear functional or functional differential equations with causal operators has preoccupied a good deal of authors. The most comprehensive sources are N.V. Azbelev, V.P. Maksimov and L.F. Rakhmatullina [1] and N.V. Azbelev and L.F. Rakhmatullina [2]. They include long lists of references, particularly concerned with the literature in Russian. The results included in this chapter have been selected from C. Corduneanu [8], Y. Li [3], [4] and M. Mahdavi [1], [2], [3]. See also Y. Li and M. Mahdavi [1]. An early noteworthy contribution is V.P. Maksimov [1].

5 Stability theory

5.1 Definition and generalities

In order to develop a theory of stability for functional differential equations with causal operators, the natural framework appears to be the theory of the second kind of initial value problems, as described in Chapter 3, Section 7 for the general case, and in Chapter 4, Section 4, for the linear case.

More precisely, we will consider the functional differential equation with causal operator

$$\dot{x}(t) = (Vx)(t), \quad t \in [0, \infty), \tag{1}$$

under initial conditions

$$x(t) = \varphi(t), \quad t \in [0, \tau), \quad x(\tau) = x^0, \tag{2}$$

and assuming that $x(t) = \theta \in \mathbb{R}^n$ is an equilibrium position for the system described by (1):

$$(V\theta)(t) = \theta \in \mathbb{R}^n, \quad t \in [0, \infty). \tag{3}$$

Under assumption of uniqueness for (1), (2), it is obvious that the null solution corresponds to the initial data $\varphi(t) = \theta, t \in [0, \tau)$, and $x(\tau) = \theta$.

The definitions for various kinds of stability will be formulated with regard to the null solution $x(t) = \theta, t \in [0, \infty)$. The case of an arbitrary solution to (1) can be reduced to the case of equilibrium by means of substitution. Indeed, if we have in mind a particular solution of (1), say $y(t)$, then substituting in (1), $x(t) = z(t) + y(t)$, and taking $z(t)$ as the new dependent variable, one obtains

$$\dot{z}(t) = (\tilde{V}z)(t), \quad t \in [0, \infty),$$

where

$$(\tilde{V}z)(t) = (V(z + y))(t) - (Vy)(t).$$

We obviously have $(\tilde{V}\theta)(t) = \theta$, which is exactly condition (3). Therefore, we can (at least theoretically!) assume that the dynamical system described by (1) has $x(t) = \theta$ as an equilibrium position.

Formulating the definition of various kinds of stability in the case of the null solution $x(t) = \theta$, will enable us to develop a stability theory similar, in many aspects, to the classical theory of stability for ordinary differential equations or for delay equations. At the present time, only a beginning has been made in regard to the stability theory of functional differential equations with causal operators, despite a large number of publications on this subject. Only special cases have been dealt with.

Let us point out that stability theory will be developed here under the basic assumption that the existence and uniqueness of the solution of the problem (1), (2) hold true. It has been made clear in Chapter 4, Section 4, that this is the case for linear equations. In Chapter 3, Section 7, we have shown that a Lipschitz type condition for the operator V, in equation (1), renders the same type of framework as the continuity of the linear operator does.

Let us now formulate the definitions of various types of stability we shall investigate in this chapter. We will assume that the underlying space for this investigation is the space $L^2_{loc}([0, \infty), \mathbb{R}^n)$, while the initial function $\varphi \in L^2_{loc}([0, \tau], \mathbb{R}^n)$. Under a Lipschitz type condition for the operator V, specific for L^2-theory, we have proven existence and uniqueness of the solution of the problem (1), (2). Of course, the exact formulation of the definition depends on the nature of the space in which the solution is sought, and the choice of space to which the initial function belongs.

We shall say that the null solution of (1) is *stable* if for every $\tau > 0$, and every $\varepsilon > 0$, there exists $\delta = \delta(\varepsilon, \tau) > 0$, such that

$$|x(t; \tau, x^0, \varphi)| < \varepsilon \quad \text{for } t \geq \tau,$$

as soon as

$$|x^0| < \delta \quad \text{and} \quad |\varphi|_2 < \delta.$$

We shall say that the zero solution of (1) is *uniformly stable* if the number $\delta(\varepsilon, \tau)$ from the definition of stability can be chosen independently of $\tau : \delta(\varepsilon, \tau) \equiv \delta(\varepsilon)$.

Obviously, uniform stability implies stability.

We shall say that the zero solution of (1) is *asymptotically stable*, if it is *stable* in the sense of the above definition, and for each $\tau > 0$ there exists $\gamma(\tau) > 0$, such that

$$\lim_{t \to \infty} |x(t; \tau, x^0, \varphi)| = 0, \tag{4}$$

for all solutions of (1) with $|x^0| < \gamma$ and $|\varphi|_2 < \gamma$.

The relationship between asymptotic stability and stability is obvious. With respect to uniform stability, it is known (see, for instance, C. Corduneanu [3], [23]) that they are independent concepts (neither one implies the other), even in the case when $(Vx)(t) = f(t, x(t))$.

In regard to condition (4), we can reformulate it in the following way: for each $\tau > 0$ and $\varepsilon > 0$, there exist $\gamma(\tau) > 0$ and $T(\varepsilon, \tau) > 0$, such that

$$|x(t; \tau, x^0, \varphi)| < \varepsilon \quad \text{for } t \geq \tau + T(\varepsilon, \tau),$$

for all x^0 and φ satisfying

$$|x^0| < \gamma(\tau), \quad |\varphi|_2 < \gamma(\tau).$$

We shall say that the zero solution of (1) is *uniformly asymptotically stable* if it is *uniformly stable* in the sense of the above definition, and if there exists $\gamma > 0$, and for each $\varepsilon > 0$ there exists $T(\varepsilon) > 0$, such that

$$|x(t; \tau, x^0, \varphi)| < \varepsilon \quad \text{for } t \geq \tau + T(\varepsilon),$$

as soon as

$$|x^0| < \gamma, \quad |\varphi|_2 < \gamma.$$

From the last definition we derive that *uniform asymptotic stability* implies both *uniform stability* and *asymptotic stability*. Therefore, it does imply all types of stabilities defined above in this section.

There is one more type of stability we shall examine in this chapter. The last definition we need to develop the stability theory of functional differential equations with causal operators can be formulated as follows:

The zero solution of (1) is *exponentially asymptotically stable* if there exist some positive constants N, α and γ, such that

$$|x(t; \tau, x^0, \varphi)| \le N[|x^0| + |\varphi|_2] \exp\{-\alpha(t - \tau)\}, \quad t \ge \tau,$$

as soon as $|x^0| < \gamma, |\varphi|_2 < \gamma$.

It is an elementary fact to check that *exponential asymptotic stability* implies *uniform asymptotic stability*, and, therefore, all types of stability defined above.

Before concluding this section, let us notice that for the solution $x(t; \tau, x^0, \varphi)$ we have always used the $\mathbb{R}^n$-norm (one of the various equivalent norms). This is motivated by the fact that all solutions of (1), in the framework of L^2-theory, are absolutely continuous on each finite interval of the semi-axis $[0, \infty)$. And for continuous functions, this norm is the most adequate, better describing the properties of the solution. (Of course, from $|x(t; \tau, x^0, \varphi)| < \varepsilon, t \ge \tau$, with $|\cdot|$ standing for the $\mathbb{R}^n$-norm, one derives $|x|_C \le \varepsilon$ in $C([0, \infty), \mathbb{R}^n)$.)

5.2 Stability of linear systems

We shall now consider the linear (homogeneous) system of functional differential equations

$$\dot{x}(t) = (Lx)(t), \quad t \in [0, \infty), \tag{5}$$

where $L : L^2_{loc}([0, \infty), \mathbb{R}^n) \to L^2_{loc}([0, \infty), \mathbb{R}^n)$ is continuous (i.e., bounded on each space $L^2([0, T], \mathbb{R}^n), 0 < T < \infty)$. By the general definition of linear operators on linear spaces we know that

$$(L\theta)(t) = \theta, \quad t \in [0, \infty). \tag{6}$$

As seen in Chapter 4, Section 4, the solution of equation (5), with initial data (4), can be represented by the formula

$$x(t; \tau, x^0, \varphi) = X(t, \tau)x^0 + \int_0^\tau \tilde{X}(t, s; \tau)\varphi(s) \, ds, \tag{7}$$

where $X(t, \tau)$ is the Cauchy matrix associated with the homogeneous system (5), and $\tilde{X}(t, s; \tau)$ has also been defined in Chapter 4, Section 2. Of course, the representation (7) is unique, and is valid for any $x^0 \in \mathbb{R}^n$ and $\varphi \in L^2([0, \tau], \mathbb{R}^n)$.

We shall now provide necessary and sufficient conditions for the stability of the zero solution of (5), in various senses defined in the preceding section. Let us examine each case we considered in the definitions.

First, we are going to prove that the *stability* of the solution $x(t) = \theta$ of (5) is equivalent to the following conditions on the matrices $X(t, s)$ and $\tilde{X}(t, s; \tau)$:

1   There exists a positive function $M(\tau)$, $\tau \geq 0$, such that

$$|X(t, \tau)| \leq M(\tau), \quad t \geq \tau. \tag{8}$$

2   For the same $M(\tau)$, one has

$$\left\{ \int_0^\tau \left| \tilde{X}(t, s; \tau) \right|^2 ds \right\}^{1/2} \leq M(\tau), \quad t \geq \tau. \tag{9}$$

We shall first prove the *sufficiency* of conditions 1 and 2 for stability. Indeed, on behalf of these conditions we obtain from formula (7)

$$|x(t; \tau, x^0, \varphi)| \leq |X(t; \tau)||x^0| + \left(\int_0^\tau \left| \tilde{X}(t, s; \tau) \right|^2 ds \right)^{1/2} \left(\int_0^\tau |\varphi(s)|^2 ds \right)^{1/2}$$

$$\leq M(\tau)[|x^0| + |\varphi|_2]$$

$$< \varepsilon \quad \text{for } t \geq \tau,$$

as soon as

$$|x^0| + |\varphi|_2 < \frac{\varepsilon}{M(\tau)} = \delta(\varepsilon, \tau).$$

By $|\varphi|_2$ we mean the norm of φ in the space $L^2([0, \tau], \mathbb{R}^n)$.

Therefore, conditions 1 and 2 are sufficient for the stability of the zero solution of (5).

Let us now prove that these conditions are also necessary.

We shall notice that the right-hand side of (7) is the sum of two solutions for (5). The first solution is

$$x(t; \tau, x^0, \theta) = X(t, \tau)x^0, \tag{10}$$

and corresponds to $\varphi = \theta$, while the second solution is

$$x(t; \tau, \theta, vf) = \int_0^\tau \tilde{X}(t, s; \tau)\varphi(s) \, ds, \tag{11}$$

corresponding to $x^0 = \theta$.

We shall proceed now in deriving (8), under the assumption that $x(t) = \theta$ is stable. If we choose $\varepsilon = 1$, there exists $\delta(1, \tau) = \delta(\tau) > 0$, such that (with τ fixed)

$$|X(t, \tau)x^0| < 1 \quad \text{for } t \geq \tau, \tag{12}$$

and all $x^0 \in \mathbb{R}$ with $|x^0| \leq \delta(\tau)$. The inequality (12) may also be written as

$$|X(t, \tau)y^0| < [\delta(\tau)]^{-1}, \quad t \geq \tau, \tag{13}$$

for all $y^0 \in \mathbb{R}^n$, with $|y^0| \leq 1$. In (13), let us choose $y^0 = e_m = $ the unit vector in the direction of the axis Ox_m. Then (13) becomes

$$|\text{col}_m X(t, \tau)| < [\delta(\tau)]^{-1}, \quad t \geq \tau. \tag{14}$$

If we use the Euclidean norm in $\mathbb{R}^n$, then (14), for $n = 1, 2, \ldots, m$ leads to

$$|X(t, \tau)| \leq \sqrt{n}[\delta(\tau)]^{-1}, \quad t \geq \tau,$$

which is nothing else but (8).

We shall now use the solution (11) in order to derive the necessity of (9). Using again the definition of stability, we can write ($\varepsilon = 1$)

$$\left| \int_0^\tau \tilde{X}(t, s; \tau)\varphi(s) \, ds \right| < 1, \quad t \geq \tau, \tag{15}$$

for all $\varphi \in L^2([0, \tau], \mathbb{R}^n)$, such that $|\varphi|_2 \leq \eta(\tau)$, with $\eta > 0$. If one fixes t and τ, $t \geq \tau$, then

$$\varphi \rightarrow \int_0^\tau \tilde{X}(t, s; \tau)\varphi(s) \, ds \tag{16}$$

is a linear map from $L^2([0, \tau], \mathbb{R}^n)$ into $\mathbb{R}^n$. Condition (15) yields, for $t \geq \tau$,

$$\left| \int_0^\tau \tilde{X}(t, s; \tau)\varphi(s)ds \right| < [\eta(\tau)]^{-1}, \tag{17}$$

for all $\varphi \in L^2([0, \tau], \mathbb{R}^n)$, such that $|\varphi|_2 \leq 1$. Once again, (17) yields for $t \geq \tau$,

$$\left| \int_0^\tau \tilde{X}(t, s; \tau)\varphi(s)ds \right| < [\eta(\tau)]^{-1}|\varphi|_2, \tag{18}$$

with $\varphi \in L^2([0, \tau], \mathbb{R}^n)$ arbitrary. At this moment in the proof, we apply a well-known result from Hilbert space theory, concerning linear continuous functionals (the Riesz representation theorem and the norm of the functional), which leads to the inequality

$$\left\{ \int_0^\tau \left| \tilde{X}(t, s; \tau) \right|^2 ds \right\}^{1/2} \leq [\eta(\tau)]^{-1}, \tag{19}$$

for $t \geq \tau$. The inequality (19) is nothing else but (9).

Of course, $M(\tau) = \max\{\sqrt{n}[\delta(\tau)]^{-1}, [\eta(\tau)]^{-1}\}$.

This ends the proof of necessity of conditions 1 and 2 for the stability of the zero solution of (5).

Remark As is the case with linear ordinary differential equations, the stability of the zero solution is equivalent to the boundedness of all solutions on $t \geq \tau$.

Let us now consider the case of *asymptotic stability* of the solution $x(t) \equiv \theta$ of (5).

Since asymptotic stability means stability plus the vanishing at ∞ of all the solutions of (5), we can claim that both conditions (8) and (9) are necessary. In fact, asymptotic stability is characterized by the validity of the following conditions:

For each $\tau \geq 0$, one has

$$\lim_{t \to \infty} |X(t, \tau)| = 0; \tag{20}$$

For each $\tau \geq 0$, one has

$$\lim_{t \to \infty} \left\{ \int_0^\tau \left| \tilde{X}(t, s; \tau) \right|^2 ds \right\}^{1/2} = 0. \tag{21}$$

As expected, condition (20) is stronger than condition (8), while condition (21) is stronger than condition (9).

It is obvious that conditions (20) and (21) are *sufficient* for asymptotic stability. This follows from formula (7).

Let us now prove the necessity of conditions (20) and (21) for asymptotic stability. As shown above, we can deal separately with solutions (10) and (11).

In the case of solution (10), one can write

$$|X(t, \tau)x^0| < \varepsilon \quad \text{for } t \geq \tau + T(\varepsilon, \tau), \tag{22}$$

provided $|x^0| < \gamma(\tau)$. Proceeding in the same manner as above (the case of stability), one derives from (22) the following estimate for the norm of $X(t, \tau)$:

$$|X(t, \tau)| < \sqrt{n}[\gamma(\tau)]^{-1}\varepsilon \quad \text{for } t \geq \tau + T(\varepsilon, \tau),$$

which immediately leads to (20).

In the case of solution (11), one again obtains as above

$$\left\{ \int_0^\tau \left| \tilde{X}(t, s; \tau) \right|^2 ds \right\}^{1/2} \leq \varepsilon \quad \text{for } t \geq \tau + T(\varepsilon, \tau),$$

which implies (9).

This shows that (20) and (21) are necessary and sufficient for asymptotic stability.

In the case of *uniform stability*, the following two conditions are necessary and sufficient. The matrix $X(t, \tau), t \geq \tau > 0$, is totally bounded, i.e., there exists $M > 0$, such that

$$|X(t, \tau)| \leq M \quad \text{for } t \geq \tau, \tag{23}$$

and

$$\left\{ \int_0^\tau \left| \tilde{X}(t, s; \tau) \right|^2 ds \right\}^{1/2} \leq M \quad \text{for } t \geq \tau. \tag{24}$$

While the sufficiency of (23) and (24) is immediate, based on formula (7) for the solution of (5), the necessity is derived as in the case of stability, noticing that $\delta(\tau)$ can be chosen to be constant (independent of $\tau \geq 0$). Similarly, $\eta(\varepsilon)$ can be chosen constant, from which assumption the estimate (21) follows.

The case of *uniform asymptotic stability* can also be treated in the same way as above, leading to the following conditions, both necessary and sufficient:

For each $\varepsilon > 0$, there exists $T(\varepsilon) > 0$, with the property that for any $\tau \geq 0$,

$$|X(t, \tau)| < \varepsilon \quad \text{for } t \geq \tau + T(\varepsilon), \tag{25}$$

and

$$\left\{ \int_0^\tau \left| \tilde{X}(t, s; \tau) \right|^2 ds \right\}^{1/2} < \varepsilon \quad \text{for } t \geq \tau + T(\varepsilon). \tag{26}$$

Finally, we shall discuss the case of *exponential asymptotic stability*. In this case, the following conditions are necessary and sufficient.

There exist two positive numbers N and α, such that

$$|X(t, \tau)| \le N \exp\{-\alpha(t - \tau)\}, \quad t \ge \tau \ge 0, \tag{27}$$

and

$$\left\{ \int_0^\tau |\tilde{X}(t, s; \tau)|^2 \, ds \right\}^{1/2} \le N \exp\{-\alpha(t - \tau)\}, \quad t \ge \tau \ge 0. \tag{28}$$

The sufficiency of conditions (27) and (28) follows easily if we use the representation (7) for the solution $x(t; \tau, x^0, \varphi)$.

The necessity of (27) and (28) is obtained following the same pattern as in the case of stability. Let us deal in detail with condition (27). According to the definition of exponential asymptotic stability (see the preceding section), one can write

$$|x(t; \tau, x^0, \theta)| = |X(t, \tau)x^0| \le N_0 |x^0| \exp\{-\alpha_0(t - \tau)\},$$

for $t \ge \tau$ and $|x^0| < \gamma = \text{const}$. The above inequality can be rewritten as

$$|X(t, \tau)x^0| \le N_0 \gamma^{-1} \exp\{-\alpha_0(t - \tau)\}, \quad t \ge \tau, \tag{29}$$

for all $x^0 \in \mathbb{R}^n$, with $|x^0| \le 1$. As above, from (29) we obtain

$$|X(t, \tau)| \le \sqrt{n} \gamma^{-1} N_0 \exp\{-\alpha_0(t - \tau)\}, \tag{30}$$

for $t \ge \tau \ge 0$, which means that estimate (27) holds true, with $N = \sqrt{n}\gamma^{-1}N_0$ and $\alpha = \alpha_0$.

A similar argument leads to the necessity of condition (28).

We shall summarize the discussion above in the following result.

Theorem 5.1   *Consider the functional differential equation (5), where $L : L_{\text{loc}}^2([0, \infty),$ $\mathbb{R}^n) \to L_{\text{loc}}^2([0, \infty), \mathbb{R}^n)$ is linear, causal and continuous. Then, the zero solution of (5) is:*

a   stable, *iff conditions (8) and (9) are satisfied*
b   uniformly stable, *iff condition (23) and (24) are satisfied*
c   asymptotically stable, *iff conditions (20) and (21) are satisfied*
d   uniformly asymptotically stable, *iff conditions (25) and (26) hold true*
e   exponentially asymptotically stable, *iff there exist positive numbers N and α, such that estimates (27) and (28) are satisfied.*

Remark 1   These characterizations of various types of stability for equation (5), under initial conditions (2), generalize those given by R. Conti for linear ordinary differential equations. See G. Sansone and R. Conti [1].

Their significance is both theoretical and practical. We will use them in investigating the stability in the first approximation, in the next section, but we will also illustrate their use in connection with special types of functional differential equations.

Remark 2   In the case of linear ordinary differential equations, uniform asymptotic stability and exponential asymptotic stability are equivalent concepts. See, for instance, C. Corduneanu [23].

A natural question arising in this respect is whether such a fact can be extended to linear functional differential equations. The answer is negative, as we shall see from the following example.

We assume that

$$(Lx)(t) = Ax(t) + \int_0^t B(t-s)x(s)\,ds,$$

where A is a constant matrix of type $n \times n$, and $B(t)$ is a matrix-valued map such that $|B(t)| \in L^1([0,\infty), \mathbb{R})$. If we consider the Laplace transform of $B(t)$,

$$\widehat{B}(s) = \int_0^\infty e^{-ts} B(t)\,dt,$$

which is defined in the semi-plane Re $s \geq 0$, then the condition

$$\det(sI - A - \widehat{B}(s)) \neq 0 \quad \text{for Re } s \geq 0$$

assures the asymptotic stability of the zero solution of $\dot{x}(t) = (Lx)(t)$. Since the operator L considered above is time invariant, the stability is actually uniform asymptotic. See C. Corduneanu [23] (Theorem 6.3.1).

S. Murakami [1] has shown that exponential asymptotic stability occurs only in case $|B(t)|\exp(\lambda t) \in L^1([0,\infty), \mathbb{R})$, for some $\lambda > 0$. Therefore, for arbitrary $B(t)$, $|B(t)| \in L^1([0,\infty), \mathbb{R})$, uniform asymptotic stability of the zero solution of (5), with $(Lx)(t)$ shown above, does not imply exponential asymptotic stability.

Remark 3   Let us consider the integrodifferential equation

$$\dot{x}(t) = \int_0^t A(t,s)x(s)\,ds, \quad t \in [0,\infty), \tag{31}$$

under initial conditions (2):

$$x(t) = \varphi(t), \quad t \in [0,\tau), \quad x(\tau) = x^0.$$

Taking into account (31) and the initial conditions (2), one can write

$$\dot{x}(t) = \int_0^\tau A(t,s)\varphi(s)\,ds + \int_\tau^t A(t,s)x(s)\,ds, \quad t \in [0,\infty), \tag{32}$$

and after integration

$$x(t) = f(t) + \int_\tau^t k(t,s)x(s)\,ds, \quad t \in [\tau,\infty), \tag{33}$$

where

$$f(t) = x^0 + \int_0^\tau \int_\tau^t A(u,s)\varphi(s)\,du\,ds, \quad t \geq \tau, \tag{34}$$

and

$$k(t, s) = \int_s^t A(u, s)\, du, \quad \tau \le s \le t < \infty. \tag{35}$$

Let us denote by $\tilde{k}(t, s)$ the resolvent kernel of $k(t, s)$, which allows us to write the solution of equation (33) in the form

$$x(t) = f(t) + \int_\tau^t \tilde{k}(t, s) f(s)\, ds, \quad t \ge \tau, \tag{36}$$

which becomes after replacing $f(t)$ by (34)

$$x(t) = \left(I + \int_\tau^t \tilde{k}(t, s)\, ds \right) x^0$$
$$+ \int_0^\tau \left\{ \int_\tau^t A(u, s)\, du + \int_\tau^t \tilde{k}(t, u) \left(\int_\tau^u A(v, s) dv \right) du \right\} \varphi(s)\, ds.$$

If we compare the last formula with (7), then we find

$$X(t, \tau) = I + \int_\tau^t \tilde{k}(t, s)\, ds, \quad \tau \le t < \infty, \tag{37}$$

and

$$\tilde{X}(t, s; \tau) = \int_\tau^t A(u, s)\, du + \int_\tau^t \tilde{k}(t, u) \left(\int_\tau^u A(v, s) dv \right) du, \tag{38}$$

the formulae (37) and (38) thus defining the kernels appearing in the representation (7). By means of the above notation, the solution $x(t) = x(t; \tau, x^0, \varphi)$ of the integrodifferential equation (31) is given by formula (7).

In order to obtain formula (7) for the solution of (31), with $X(t, \tau)$ and $\tilde{X}(t, s; \tau)$ given by (37) and (38), we have performed some operations which must be justified. There are many possibilities (i.e., conditions imposed on $A(t, s)$) to justify the validity of the above operations.

For intance, if we want to use the same framework as above (in this section), which basically means L^2-theory, then the following assumptions on $A(t, s)$ and $\varphi(s)$ are in order.

1   $A(t, s)$ satisfies the Hilbert–Schmidt condition

$$\int_0^t |A(t, s)|^2 ds \in L^\infty_{\text{loc}}([0, \infty), \mathbb{R}); \tag{39}$$

2   $\varphi \in L^2([0, \tau], \mathbb{R}^n)$.

Under these assumptions, there exists $\tilde{k}(t, s)$ satisfying a condition similar to (39) and all operations performed above to determine $X(t, \tau)$ and $\tilde{X}(t, s; \tau)$ are valid.

Let us now provide conditions, based on Theorem 5.1, to assure the *uniform stability* of the zero solution of equation (31). We must provide conditions such that (23) and (24) are both satisfied.

Taking into account (37), it is obvious that (23) is satisfied provided

$$\text{ess-sup} \left\{ \int_0^t \left| \tilde{k}(t, s) \right| ds, t \geq 0 \right\} = M < \infty. \tag{40}$$

Of course, we would like to have a condition on $k(t, s)$ given by (35), instead of (40). This is due to the fact that $\tilde{k}(t, s)$ is not always easy to construct (and then to check (40)). See Lemma 4.2.2 in C. Corduneanu [10], in which a condition is provided (a result due to O. Staffans) for the validity of (40). This condition, imposed to $A(t, s)$, looks like

$$\sup \left\{ \int_0^t \left(\int_s^t |A(u, s)| \, du \right) ds; \ t \geq 0 \right\} < \infty. \tag{41}$$

We have to make such assumptions on $A(t, s)$ that will lead to the boundedness of $|\tilde{X}(t, s; \tau)|$, which means condition (24). Once condition (40) has been accepted, it is obvious from (38) that we need to impose the boundedness of $\int_0^t A(u, s) \, du$ in the angle $\Delta = \{(t, s); \ 0 \leq s \leq t < \infty\}$, in order to derive the boundedness of $\tilde{X}(t, s; \tau)$. Hence, we will add the assumption

$$\sup \left\{ \int_0^t |A(u, s)| \, du; \ (t, s) \in \Delta \right\} < \infty. \tag{42}$$

We find that $A(t, s)$ must be subject to conditions (39), (41) and (42). It is natural to ask whether such kernels exist.

It is easy to check that any $A(t, s)$ defined on Δ, such that

$$|A(t, s)| \leq a(t)b(s), \tag{43}$$

with $a(t)$ and $b(s)$ satisfying $a, b \in L^1([0, \infty), \mathbb{R}) \cap L^\infty([0, \infty), \mathbb{R}^n)$, is acceptable. There are many choices for $a(t)$ and $b(s)$. Let us notice that both $a(t)$ and $b(s)$ belong to any $L^p([0, \infty), \mathbb{R})$, $1 \leq p \leq \infty$, in particular, $b \in L^2$ being necessary in (39).

5.3 Stability of quasilinear systems

Using the result obtained in the preceding section, one can investigate the stability of certain nonlinear systems. In particular, the systems with a "weak" nonlinearity are adequate to this kind of investigation. In stability theory (of ordinary differential equations), one also speaks about "stability in the first approximation", or stability of perturbed systems.

More precisely, we will consider systems of the form

$$\dot{x}(t) = (Lx)(t) + (Fx)(t), \quad t \in [0, \infty), \tag{44}$$

in which L stands for a linear operator of causal type, continuous on a given underlying space.

Assuming that the zero solution of the linear part of the system (44)

$$\dot{x}(t) = (Lx)(t), \quad t \in [0, \infty),$$

which is exactly equation (5), has a certain kind of stability, we will seek conditions for the "nonlinearity" $(Fx)(t)$, such that the zero solution of (44) will also be stable.

A rather elementary result is concerned with *uniform stability* of the zero solution.

Theorem 5.2 *Consider the system* (44) *under the following assumptions:*

1   $L : L^2_{loc}([0, \infty), \mathbb{R}^n) \to L^2_{loc}([0, \infty), \mathbb{R}^n)$ *is a linear, causal and continuous operator.*
2   *The solution $x(t) = \theta$ of* (5) *is uniformly stable.*
3   $F : L^2_{loc}([0, \infty), \mathbb{R}^n) \to L^2_{loc}([0, \infty), \mathbb{R}^n)$ *is a continuous causal operator, such that*

$$|(Fx)(t)| \le \gamma(t)|x(t)| \quad a.e. \text{ on } [0, \infty).$$

Then, the zero solution of (44) *is also uniformly stable.*

Proof   As we know, the solution of the system $\dot{x}(t) = (Lx)(t) + f(t)$, under initial conditions (4), can be represented by the formula

$$x(t) = X(t, \tau)x^0 + \int_0^\tau \tilde{X}(t, s; \tau)\varphi(s)\, ds + \int_\tau^t X(t, s)f(s)\, ds.$$

Of course, it is assumed that $\varphi \in L^2([0, \tau], \mathbb{R}^n)$ and $f \in L^2_{loc}([0, \infty), \mathbb{R}^n)$.

Using the above formula for the solution, we can write the following functional integral equation for the solution of (44):

$$x(t; \tau, x^0, \varphi) = X(t, s)x^0 + \int_0^\tau \tilde{X}(t, s; \tau)\varphi(s)\, ds + \int_\tau^t X(t, s)(Fx)(s)\, ds. \tag{45}$$

Therefore, if there exists a solution of (44) with initial conditions (4), then this solution must satisfy a.e. (45), on its interval of existence. Viceversa, a solution of (45), will satisfy a.e. on its existence interval the equation (44). The existence interval is always of the form $[\tau, T), T \le \infty$.

Since $x(t) = \theta$ is uniformly stable for (5), the conditions (23) and (24) must hold. Therefore, equation (45) leads to the following integral inequality:

$$|x(t; \tau, x^0, \varphi)| \le M(|x^0| + |\varphi|_2) + M \int_\tau^t \gamma(s)|x(s; \tau, x^0, \varphi)|\, ds, \quad t \ge \tau. \tag{46}$$

The inequality (46) is of Gronwall type and is valid on the interval of existence of the solution $x(t; \tau, x^0, \varphi)$ – which interval may not coincide with the semi-axis $t \ge \tau$.

One derives from (46) the following estimate for $|x(t; \tau, x^0, \varphi)|$:

$$|x(t; \tau, x^0, \varphi)| \le M(|x^0| + |\varphi|_2)\exp\left\{ M \int_0^\infty \gamma(s)\, ds \right\},$$

which can also be written as

$$|x(t; \tau, x^0, \varphi)| \le \bar{M}(|x^0| + |\varphi|_2), \quad t \ge \tau, \tag{47}$$

with $\overline{M} = M \exp\{M \int_0^\infty \gamma(s)\, ds\}$.

The estimate (47) is valid for $t \ge \tau$, as long as the solution $x(t; \tau, x^0, \varphi)$ exists in the future. But the fact that $|x(t; \tau, x^0, \varphi)|$ remains bounded on its existence interval, which implies the boundedness in $L^\infty \oplus L^1$ of its derivative on the same interval, shows that $\lim x(t; \tau, x^0, \varphi)$ exists as $t \to T$ = the right extremity of the interval of existence of the solution. Therefore, the solution can be extended beyond T, which means that the only possibility is $T = \infty$. With (47) valid on $[\tau, \infty)$, the uniform stability of the zero solution of (44) is thereby proven.

Remark   The growth condition on $(Fx)(t)$ in condition 3 above could be somewhat weakened. Indeed, if we replace the inequality in condition 3 by

$$|(Fx)(t)| \le \gamma(t) \sup_{0 \le s \le t} |x(s)| \quad \text{a.e. on } [0, \infty),$$

which makes sense only for continuous $x(t)$ (or $x \in L^\infty$, with ess-sup instead of sup), then we still get the same result. We point out the feature that any time we have used the inequality of condition 3, we dealt with an $x(t)$ which is a solution of (44), hence, absolutely continuous.

The above case can be handled in exactly the same way as in the proof of Theorem 5.2, if we notice that the right-hand side of (46) is a nondecreasing function of t, $t \ge \tau$.

We shall now examine the case of exponential asymptotic stability for the zero solution of equation (44).

Theorem 5.3   *Consider equation (44) under the following assumptions:*

1   *the same as in Theorem 5.2;*
2   *the solution $x(t) = \theta$ if (5) is exponentially asymptotically stable;*
3   *F is a continuous operator on $L^2_{\text{loc}}([0, \infty), \mathbb{R}^n)$, and satisfies the inequality*

$$|(Fx)(t)| \le \mu |x(t)| \quad \text{a.e. on } [0, \infty),$$

for each continuous x, with $\mu > 0$ sufficiently small.

Then, the solution $x(t) = \theta$ of (44) is also exponentially asymptotically stable.

Proof   We proceed exactly along the same lines as in the proof of Theorem 5.2. The solution $x(t; \tau, x^0, \varphi)$ satisfies the functional integral equation (45). Since the solution $x(t) = \theta$ of (5) is exponentially asymptotically stable, conditions (27) and (28) are satisfied. One obtains, also taking into account condition 3 in the theorem, the following integral inequality:

$$|x(t; \tau, x^0, \varphi)| \le N(|x^0| + |\varphi|_2) \exp\{-\alpha(t - \tau)\}$$

$$+ N\mu \exp\{-\alpha t\} \int_\tau^t \exp\{\alpha s\} |x(s; \tau, x^0, \varphi)| \, ds.$$

We multiply both sides of the above inequality by $\exp\{\alpha t\}$, and denote $u(t) = \exp\{\alpha t\}|x(t; \tau, x^0, \varphi)|$. One obtains the Gronwall type inequality

$$u(t) \le N(|x^0| + |\varphi|_2) \exp\{\alpha \tau\} + N\mu \int_\tau^t u(s) \, ds, \tag{48}$$

valid on the interval of existence of the solution $x(t; \tau, x^0, \varphi)$. Inequality (48) yields

$$|x(t; \tau, x^0, \varphi)| \le N(|x^0| + |\varphi|_2) \exp\{(N\mu - \alpha)(t - \tau)\}.$$

Let us now assume that μ is chosen such that $N\mu - \alpha = -\alpha_0 < 0$. Therefore,

$$|x(t; \tau, x^0, \varphi)| \leq N(|x^0| + |\varphi|_2) \exp\{-\alpha_0(t - \tau)\}, \tag{49}$$

which proves the exponential asymptotic stability of the zero solution of (44).

The fact that the solution is defined on $[0, \infty)$ results in the same way as in the proof of Theorem 5.2.

Remark 1   Again, as in case of Theorem 5.2, we notice that the condition 3 is rather restrictive. Indeed, the left-hand side of the inequality is a causal operator, while the estimate is typical for functions: $(Fx)(t) = f(t, x(t))$, i.e., Niemytskii type operators (very special, in comparison with causal operators).

In fact, condition 3 is not as restrictive as may appear at first glance. For instance, F may be chosen in the form

$$(Fx)(t) = (F_0 x)(t) \cdot x(t),$$

where F_0 stands for a matrix-valued map whose elements are causal operators. If F_0 is small enough in norm, it is clear that the corresponding $(Fx)(t)$ will satisfy a condition of the form stipulated in condition 3 of Theorem 5.2. An immediate example is

$$(Fx)(t) = A \left(m + \int_0^t |x(s)| \, ds \right)^{-1} x(t),$$

where A is a n by m matrix, and $m > 0$ is large enough.

Remark 2   More can be said about the solutions of (44) if condition 3 in Theorem 5.3 is replaced by (the stronger condition)

$$|(Fx)(t) - (Fy)(t)| \leq \mu |x(t) - y(t)|, \tag{50}$$

while assuming

$$|(F\theta)(t)| \in L^\infty([0, \infty), \mathbb{R}). \tag{51}$$

Then, from (45) one obtains the integral inequality

$$|x(t)| \leq K(|x^0| + |\varphi|_2) \exp\{-\lambda(t - \tau)\}$$
$$+ L \int_\tau^t [\mu |x(s)| + |(F\theta)(s)|] \exp\{-\lambda(t - s)\} \, ds,$$

with K, L and λ positive constants.

Using a result of A.A. Martynyuk [1], the above inequality yields for $t \geq \tau$

$$|x(t)| \leq K(|x^0| + |\varphi|_2) \exp\{-\lambda(t - \tau)\} + \frac{A}{\lambda - \mu},$$

admitting, of course, that $\mu < \lambda$.

Hence, any solution of (44) is bounded on $[0, \infty)$. Moreover, if $x(t)$ and $y(t)$ are two solutions of (44), corresponding to (x^0, φ), resp. (y^0, ψ), then the following estimate holds true for their difference:

$$|x(t) - y(t)| \leq K(|x^0 - y^0| + |\varphi - \psi|_2) \exp\{-(\lambda - \mu)(t - \tau)\}.$$

This estimate shows that each solution of (44) is exponentially asymptotically stable. All solutions exist on $[0, \infty)$ because the right-hand side in (44) satisfies a (global) Lipschitz condition.

5.4 Comparison method in stability

This method is a generalization of the classical method in stability theory, as this theory was shaped by A.M. Lyapunov at the end of the 1800s. Instead of trying to estimate $|x(t)|$ directly, one estimates a certain functional $(Vx)(t)$, subject to some conditions, and then one passes the information obtained for $(Vx)(t)$ onto $|x(t)|$. The comparison method relies closely on the case of differential inequalities, which allows one to obtain the necessary estimates.

We shall consider the functional differential equation (1)

$$\dot{x}(t) = (Vx)(t), \quad t \in [0, \infty),$$

with V a causal continuous operator on the space $L^2_{\text{loc}}([0, \infty), \mathbb{R}^n)$. The general existence result for equation (1), under conditions specific to the second kind of initial value problem (2), has been discussed in Chapter 3, Section 7. This was done under the assumptions of L^2-theory. Similar results can be obtained in the framework of L^1-theory (see C. Corduneanu [6]). Other cases can also be treated, following a similar procedure.

In the case of L^2-theory we want to use as backgroud the following basic assumption on V: $(V\theta)(t) = 0, t \geq 0$, as well as the Lipschitz type condition (3.145)

$$\int_0^t |(Vx)(s) - (Vy)(s)|^2 ds \leq L(t) \int_0^t |x(s) - y(s)|^2 ds,$$

where $L(t)$ is a nondecreasing function on $[0, \infty)$. We will also assume that the initial function ϕ belongs to the space L^2.

These assumptions assure the existence and uniqueness of the solution $x(t; \tau, x^0, \varphi)$, which satisfies (1) a.e. on $[\tau, \infty)$, and the initial conditions (2),

$$x(\tau; \tau, x^0, \varphi) = x^0, \quad x(s; \tau, x^0, \varphi) = \varphi(s), \quad s \in [0, \tau).$$

Actually, $x(t; \tau, x^0, \varphi)$ is defined only for $t \geq \tau$, but the second initial condition above can be regarded as completing the definition of $x(t; \tau, x^0, \varphi)$.

We will consider in this section some functionals, which we shall call *Lyapunov functionals*, $W = (Wx)(t), t \geq 0$, such that $W : L^2_{\text{loc}}([0, \infty), \mathbb{R}^n) \to \mathbb{R}$, assuming also that W is causal.

A basic assumption on W will be its Fréchet differentiability at any $x \in L^2_{\text{loc}}([0, \infty), \mathbb{R}^n)$. The Fréchet differential of the functional W at x will be denoted by $(W'x)(t)$. Since $L^2_{\text{loc}}([0, \infty), \mathbb{R}^n)$ is not a normed space, we will need to use the Fréchet space (see Chapter 2, Section 2.1) structure in order to define the Fréchet differential. This is possible, and moreover, the "chain rule" is valid (see, for instance, G. Marinescu [1]).

If $x = x(t)$ is a solution of (1), then the following basic formula holds true:

$$\frac{d}{dt}(Wx)(t) = (W'x)(t)[(Vx)(t)],\tag{52}$$

which can be read as follows: the derivative of $(Wx)(t)$, when $x(t)$ is a trajectory of (1), is obtained by taking the value of the differential operator W' at $x(t)$, in the direction $(Vx)(t)$.

The above considerations represent a generalization of the well-known fact for $(Wx)(t) = W(x_1(t), \ldots, x_n(t))$, when (52) becomes

$$\frac{d}{dt}(Wx)(t) = (\text{grad } W_x)(t) \cdot (Vx)(t).\tag{53}$$

This formula is termed, in stability theory, the formula for the derivative of W, with respect to the system (1).

The *comparison method* in stability theory of functional differential equations is based on the finding of an inequality of the form

$$(W'x)(t)[(Vx)(t)] \le \omega(t, (Wx)(t)),\tag{54}$$

for all $x \in L^2_{\text{loc}}([0, \infty), \mathbb{R}^n)$, and of the stability properties of the *comparison equation* associated with (54):

$$y'(t) = \omega(t, y(t)),\tag{55}$$

in which $\omega(t, y)$ is a scalar function for $t \ge 0$ and $y \in \mathbb{R}$. Actually, since only nonnegative Lyapunov functionals will be considered, the function $\omega(t, y)$ will be defined only in the quadrant $t \ge 0$, $y \ge 0$, and the *assumption* $\omega(t, 0) = 0$ *is needed* (so that the comparison equation (55) admits the zero solution on $t \ge 0$).

At this point in our discussion one may ask why the comparison equation (55) is an ordinary differential equation and not a functional differential equation with causal operator. The answer to this question is that we can handle in an easier manner an ordinary differential equation than a functional differential equation with causal operator, and the differential inequalities involving causal operators require a monotonicity property for the operator – a feature which is not in harmony with the stability needs (for the zero solution of the comparison equation).

In what follows, we shall rely on the possibility of finding for the Lyapunov functional W an inequality of the form (54). Using formula (53), the inequality (54) can be rewritten as

$$\frac{d}{dt}(Wx)(t) \le \omega(t, (Wx)(t)),\tag{56}$$

which is directly related to the comparison equation (55).

Based on the theory of (scalar) differential inequalities (see, for instance, C. Corduneanu [23]), one can derive from (55) and (56) inequalities of the form

$$(Wx)(t) \le y(t; \tau, y_0), \quad t \ge \tau,\tag{57}$$

provided

$$(Wx)(\tau) \le y_0,\tag{58}$$

where

$$(Wx)(\tau) = \lim_{t \to \tau-} (Wx)(t).$$

In Chapter 3, Section 7, we have proven the continuous dependence of solution of (1) with respect to the data (Theorem 3.17). Using the estimate (3.154), we can write

$$|(Wx)(\tau)| \le M(|x^0| + |\varphi|_2), \tag{59}$$

where M may depend on τ. Therefore, we can strengthen (58) by requiring

$$M(|x^0| + |\varphi|_2) \le y_0. \tag{60}$$

We shall now formulate a definition related to Lyaponov functionals, which will be used in proving stability results.

Property A   The functional $W : L_{\mathrm{loc}}([0, \infty), \mathbb{R}^n) \to [0, \infty)$, with $(W\theta)(t) = 0$ on $[0, \infty)$, is said to possess property A, if for each $\varepsilon > 0$, there exists $\delta = \delta(\varepsilon, \tau) > 0$, such that

$$[(Wx)(t) < \delta \text{ for } t \ge \tau] \Longrightarrow [|x(t)| < \varepsilon \text{ for } t \ge \tau].$$

Property A can be compared with the well-known property of positive definiteness. Indeed, if we take the classical definition for W of being positive definite, then we should be able to get a positive nondecreasing function $a(\tau, r), r \ge 0, a(\tau, 0) = 0$ in r, such that $a(t, |x(t)|) \le (Wx)(t)$. In this case, one can obviously take $\delta(\varepsilon, \tau) = a(\tau, \varepsilon)$, the property A being satisfied.

We think that the condition of positive definiteness, under formulation A, is more adequate for causal functionals.

Theorem 5.4   *Consider the functional differential equation* (1), *with* $V : L^2_{\mathrm{loc}}([0, \infty), \mathbb{R}^n) \to L^2_{\mathrm{loc}}([0, \infty), \mathbb{R}^n)$ *a causal continuous operator. Assume there exists a Fréchet differentiable Lyapunov functional* $W : L^2_{\mathrm{loc}}([0, \infty), \mathbb{R}^n) \to \mathbb{R}$, *with* $(W\theta)(t) = 0$ *for* $t \in [0, \infty)$, *satisfying the comparison inequality* (56). *Moreover, let* $\omega(t, y), \omega(t, 0) = 0, t \in [0, \infty)$, *be such that the zero solution of the comparison equation* (55) *is stable (asymptotically stable). If* W *satisfies property A, then the zero solution of* (1) *is stable (asymptotically stable).*

Proof   Let us first consider the case of stability. For each $\varepsilon > 0$ and $\tau > 0$, arbitrarily chosen, there exist $\delta = \delta(\varepsilon, \tau) > 0$ such that

$$y(t; \tau, y^0) < \delta \quad \text{for } t \ge \tau, \tag{61}$$

provided $y_0 < \eta(\varepsilon, \tau)$, with $\eta > 0$. Consequently, from (57)–(59) and (60) there results

$$(Wx)(t) < \delta \quad \text{for } t \ge \tau, \tag{62}$$

as soon as x^0 and φ are such that

$$M(\tau)(|x^0| + |\varphi|_2) < \eta. \tag{63}$$

Let us notice that inequality (63) is satisfied, provided one chooses

$$|x^0|, |\varphi|_2 < \eta(2M)^{-1}, \tag{64}$$

where $\eta(2M)^{-1}$ depends on ε and τ.

In accordance with property A and the choice of $\delta = \delta(\varepsilon, \tau)$, inequality (62) implies

$$|x(t)| < \varepsilon \quad \text{for } t \geq \tau.$$

This proves that the solution $x(t) = \theta$ of (1) is stable.

In regard to asymptotic stability, which means stability and the vanishing of solutions at infinity, we notice that in the case of asymptotic stability of the zero solution of the comparison equation (55), besides (62), inequality (57) also implies

$$\lim (Wx)(t) = 0 \quad \text{as } t \to \infty, \tag{65}$$

provided $|x^0|$ and $|\varphi|_2$ are small enough. Therefore, using (65), we can write

$$(Wx)(t) < \delta(\varepsilon', T) \quad \text{for } t \geq T \geq \tau, \tag{66}$$

as soon as $|x^0|$ and $|\varphi|_2$ are chosen sufficiently small. Now, property A and (66) imply

$$|x(t)| < \varepsilon' \quad \text{for } t \geq T,$$

which tells us that

$$\lim x(t; \tau, x^0, \varphi) = 0 \quad \text{as } t \to \infty.$$

This ends the proof of Theorem 5.4.

Remark Theorem 5.4 contains as special cases various stability results with a classic flavor. For instance, the case $\omega(t, y) = 0$ leads to the following criterion:

If there exists a Lyapunov functional W for the system (1), such that

$$(W'x)(t)[(Vx)(t)] \leq 0 \quad \text{for } t \in [0, \infty), \tag{67}$$

and all $x \in L^2_{\text{loc}}([0, \infty), \mathbb{R}^n)$, then property A has the stability of the zero solution of (1) as a consequence.

Let us notice that the meaning of (67) is as follows: W' is a linear continuous functional on $L^2_{\text{loc}}([0, \infty), \mathbb{R}^n)$, which in (67) is taken at $Vx \in L^2_{\text{loc}}([0, \infty), \mathbb{R}^n)$. The result is, of course, a real-valued function that is defined for $t \in [0, \infty)$. Hence, it may be adequate to request the validity of (67) only a.e.

Let us also notice that (67) involves $x(t)$, which is usually locally absolutely continuous (because it is a solution of (1)).

The use of a Lyapunov functional for functional differential equations is not an easy matter. We shall briefly discuss here an example due to J.J. Levin [1], which is presented in detail in C. Corduneanu [3].

Let us consider the integrodifferential equation

$$\dot{x}(t) = -\int_0^t k(t, s)\varphi(x(s))\,ds, \quad t \in [0, \infty), \tag{68}$$

all functions involved being scalar ($n = 1$) and real. The classical initial condition $x(0) = x^0 \in \mathbb{R}$ is associated with (68).

The following hypotheses are necessary for the construction of a Lyapunov functional, in the form

$$E(t) = \Phi(x(t)) + \frac{1}{2} k(t, 0) \int_0^t \varphi(x(t))\,ds + \frac{1}{2} \int_0^t k_s(t, s) \left(\int_s^t (x(u))\,du \right)^2 ds, \tag{69}$$

where

$$\Phi(u) = \int_0^u \varphi(v)\,dv, \quad u \in \mathbb{R}. \tag{70}$$

H$_1$ The kernel $k(t, s)$ is continuous on the set $\Delta = \{(t, s); \ 0 \le s \le t < \infty\}$, and continuously differentiable of the third order on the set $\tilde{\Delta} = \{(t, s); \ 0 \le s < t < \infty\}$, such that

$$k(t, s) \ge 0, \quad k_t(t, s) \le 0, \quad k_{tt}(t, s) \ge 0,$$

$$k_s(t, s) \ge 0, \quad k_{ts}(t, s) \le 0, \quad k_{tts}(t, s) \ge 0,$$

with subscripts indicating differentiation.

H$_2$ For every $f(t)$ continuous on $[0, \infty)$, the following formulas hold:

$$\frac{d}{dt}\int_0^t k(t, s) f(s)\,ds = k(t, t) f(t) + \int_0^t k_t(t, s) f(s)\,ds,$$

$$\frac{d}{dt}\int_0^t k_s(t, s)\left[\int_s^t f(u)\,du\right]^2 ds = \int_0^t k_{ts}(t, s)\left[\int_s^t f(u)\,ds\right]^2 ds$$

$$- 2k(t, 0) f(t) \int_0^t f(s)\,ds$$

$$+ 2f(t) \int_0^t k(t, s) f(s)\,ds,$$

as well as similar formulas with $k_{ts}(t, s)$ instead of $k_s(t, s)$.

Remark   The above formulas (in H$_2$) are not generally satisfied under hypothesis H$_1$, since we have assumed the smoothness of $k(t, s)$ in $\tilde{\Delta}$ only.

H$_3$ The following inequalities hold:

$$\sup_{0 \le t < \infty} k(t, t) < \infty, \quad \inf_{0 \le t < \infty} \int_0^t k_t(t, s)\,ds > -\infty,$$

$$\limsup_{t \to \infty} \int_{t-\delta}^t (t - s)^2 k_{ts}(t, s)\,ds < 0,$$

for every $\delta > 0$.

H$_4$ $\varphi(x)$ is continuous on $\mathbb{R}$, $x\varphi(x) > 0$ for $x \ne 0$, and

$$\Phi(u) \to \infty \quad \text{as } |u| \to \infty.$$

This means $\varphi(0) = 0$, which implies $x = 0$ is a solution of (68).

Theorem 5.5   *Under hypotheses* H$_1$–H$_4$, *for each* $x^0 \in \mathbb{R}$ *there exists at least one solution* $x(t)$ *of equation* (68), *defined on* $[0, \infty)$, *such that* $x(0) = x^0$ *and*

$$\lim_{t \to \infty} x^{(j)}(t) = 0, \quad j = 0, 1, 2.$$

The details of the proof are rather technical, but elementary in nature. As mentioned above, the full proof of Theorem 5.5 can be found in J.J. Levin [1] or in C. Corduneanu [3].

Let us notice that the hypotheses H_1–H_4 do not allow us to be in the framework resulting from the general assumptions, in the case of the operator

$$(Vx)(t) = -\int_0^t k(t, s)\varphi(x(s))\, ds$$

occuring in (68). These hypotheses are more appropriate for the case when V is acting on the space of continuous functions.

The reason we have discussed this equation is to illustrate the intricacies of Lyapunov's method in the case of integrodifferential equations (formally, a special case of (1)).

5.5 Admissibility concepts

The admissibility concepts are related to that of stability in various senses. Roughly speaking we say that the pair of function spaces (B, D) is *admissible* with respect to the equation $x(t) = (Lx)(t) + f(t)$, if this equation has its solution in the space D, for each $f \in B$. Nonlinear formulations can also be given for this concept. If, in particular, one chooses $D = C_0([0, \infty), \mathbb{R}^n)$, where C_0 is the subspace of C consisting of those functions vanishing at infinity, it is obvious that we obtain stability conditions (in the classical sense).

The admissibility can also be defined with respect to an operator, say T, and it consists in demanding the inclusion $TB \subset D$. For instance, the equation $x(t) = (Lx)(t) + f(t)$ can be written as $x(t) = ((I - L)^{-1}f)(t)$, which means that admissibility with respect to this equation is the same as admissibility of the pair of spaces (B, D), with respect to the operator $(I - L)^{-1} = T$, of course, assuming the existence of the inverse, as well as certain properties, like continuity.

Several results concerning the admissibility theory for classical Volterra equations have been proven in C. Corduneanu [3], [10]. In this section we will relate such results with the theory of equations with causal operators.

Admissibility theory for ordinary differential equations was started by P.G. Bohl early the twentieth century, consolidated by O. Perron in the 1930s, by R. Bellman in the 1940s, and brought to a developed stage by J.L. Massera and J.J. Schäffer in the 1950s. For a complete account of admissibility theory in the theory of ordinary differential equations, up to the mid-1960s, the reader is refered to the book of J.L. Massera and J.J. Schäffer [1].

In the theory of integral equations of Volterra type, the admissibility concept was investigated by C. Corduneanu [1] in the mid-1960s, and carried out by J.M. Cushing [1], [2], who built up an abstract theory of admissibility (with particular application to linear integral equations).

We shall not attempt now to provide a comprehensive presentation of admissibility theory. Our intention is to give an introduction to the subject, relying on the theory of equations with causal operators as developed in this book. Also, we will try to connect this theory with known results, especially with regard to (classical) integral equations.

The kind of functional equation we shall investigate in this section has the form

$$x(t) = (Lx)(t) + (Nx)(t), \quad t \in [0, \infty), \tag{71}$$

which has been considered earlier (see Chapter 4, Section 5) when we dealt with quasilinear equations (on a compact interval of $\mathbb{R}$). This time we shall be concerned with the case of functional equations, instead of functional differential equations. The latter can be reduced by integration to the form (71).

The linear counterpart of (71) is

$$x(t) = (Lx)(t) + f(t), \quad t \in [0, \infty), \tag{72}$$

where $L : L^1_{\text{loc}}([0, \infty), \mathbb{R}^n) \to L^1_{\text{loc}}([0, \infty), \mathbb{R}^n)$ is a linear, causal and continuous operator.

The first basic assumption on (72) will be the admissibility of the pair (B, D), where B and D are Banach spaces consisting of measuable functions. This means that for each $f \in B$, the unique solution x of (72) is in D.

We further assume that B and D are endowed with topologies stronger than the topology of $L^1_{\text{loc}}([0, \infty), \mathbb{R}^n)$. In other words, if $f^{(m)} \to f$ in B, then $f^{(m)} \to f$ also in $L^1_{\text{loc}}([0, \infty), \mathbb{R}^n)$.

It is simple to note the fact that the map $f \to x$, with x the solution of (72), is linear and continuous from B to D. Indeed, let $f^{(m)} \to f$ in B, as $m \to \infty$. This implies the convergence of $f^{(m)}$ to f in L^1_{loc}. Since $x^{(m)}$, the solution of (72) for $f = f^{(m)}$, can be represented by the resolvent formula

$$x^{(m)}(t) = f^{(m)}(t) + (Rf^{(m)})(t), \tag{73}$$

with R the resolvent kernel associated with L, which we know to exist as a continuous, linear and causal operator on L^1_{loc}, it follows from (73) that $x^{(m)} \to x$ in L^1_{loc}, as $m \to \infty$, with x the solution of (72). This means that the graph of the mapping $f \to x$ is closed in $B \times D$. Now, if we assume $(f^{(m)}, x^{(m)}) \to (f, \tilde{x})$ in $B \times D$, which implies $x^{(m)} \to \tilde{x}$ in D as $m \to \infty$, using the fact that convergence in D has as a consequence convergence in L^1_{loc}, we get $\tilde{x} = x$.

Based on the closed graph theorem (see, for instance, R.E. Edwards [1]), we conclude that the map $f \to x$ is continuous. Hence, there exists $K > 0$, such that

$$|x|_D \leq K|f|_B, \quad f \in B. \tag{74}$$

We shall use this partial result about admissibility to obtain the existence and uniqueness of the solution to equation (71).

Theorem 5.6 *Consider equation (71), with $L : L^1_{\text{loc}}([0, \infty), \mathbb{R}^n) \to L^1_{\text{loc}}([0, \infty), \mathbb{R}^n)$ a linear, causal and continuous operator. Assume that the pair of spaces (B, D), with topologies stronger than the topology of $L^1_{\text{loc}}([0, \infty), \mathbb{R}^n)$, is admissible with respect to the linear equation (72).*

Moreover, let $N : D \to B$ be a (nonlinear) map satisfying

$$|Nx - Ny|_B \leq \lambda |x - y|_D, \tag{75}$$

with sufficiently small λ.

Then, there exists a unique solution $x \in D$ of equation (71).

Proof   We will apply the contraction mapping principle in the space D, to the operator $u \to Tu = x$, where x is the solution (in D) of the linear equation

$$x(t) = (Lx)(t) + (Nu)(t), \quad t \in [0, \infty). \tag{76}$$

If we choose another function $v \in D$, and denote by y the solution of the equation similar to (76)

$$y(t) = (Ly)(t) + (Nv)(t), \quad t \in [0, \infty), \tag{77}$$

then for $z(t) = x(t) - y(t)$ one has

$$z(t) = (Lz)(t) + (Nu)(t) - (Nv)(t), \tag{78}$$

which, using (74), leads to

$$|z|_D \leq K |(Nu)(t) - (Nv)(t)|_B. \tag{79}$$

Taking now (75) into account we obtain from (79) the inequality

$$|(Tu)(t) - (Tv)(t)|_D \leq \lambda K |u(t) - v(t)|_D. \tag{80}$$

Consequently, if one assumes

$$\lambda < K^{-1}, \tag{81}$$

then T is a contraction map on D, which means that there exists a unique fixed element for this operator (in D). If we let $u = x$ in (76), we obtain for $x(t)$ the equation (71).

This ends the proof of Theorem 5.6.

From Theorem 5.6 it is possible to derive many existence results, by particularizing the spaces B and D, as well as the linear operator $(Lx)(t)$ which appears in equation (71).

Let us consider the case $B = D = L^\infty([0, \infty), \mathbb{R}^n)$, consisting of all measurable maps from $[0, \infty)$ into $\mathbb{R}^n$ which are essentially bounded. The usual L^∞-norm is defined by the ess-sup. We will obtain in this way an existence result for essentially bounded solutions on $[0, \infty)$.

The property of boundedness of solutions is also called *Lagrange stability* in the theory of dynamical systems.

Let us now choose the operator L in the form

$$(Lx)(t) = \int_0^t k(t, s) x(s) \, ds, \quad t \in [0, \infty), \tag{82}$$

where $k(t, s)$ is a measurable kernel, integrable in s for almost all $t \in [0, \infty)$, such that

$$\text{ess-sup} \int_0^t |k(t, s)| \, ds \leq M < \infty, \quad t \in [0, \infty). \tag{83}$$

Then the pair (L^∞, L^∞) is admissible with respect to the operator L defined by (82). Indeed, if $k(t, s)$ satisfies (83), then its resolvent kernel $\tilde{k}(t, s)$ (see Lemma 4.2.2 in C. Corduneanu [10]) satisfies a similar estimate. Since the solution of (72) can be expressed by the resolvent formula

$$x(t) = f(t) + \int_0^t \tilde{k}(t, s) f(s) \, ds,$$

it is obvious that the pair (L^∞, L^∞) is admissible with respect to the operator L defined by (82), if and only if (83) holds.

In regard to the nonlinear term N in (71), we have to assume (75), which becomes in this case

$$\text{ess-sup}|(Nx)(t) - (Ny)(t)| \leq \lambda \ \text{ess-sup}|x(t) - y(t)|,$$

the ess-sup being taken for $t \in [0, \infty)$.

Of course, one may also particularize N, taking for instance

$$(Nx)(t) = \int_0^\infty m(t, s, x(s)) \, ds,$$

under suitable conditions for m. We leave to the reader the task of formulating exact conditions on $m(t, s, x)$, such that (75) holds. We point out the fact that N need not be a causal operator on L^∞.

6   Neutral functional equations

6.1 The concept of neutral functional equation with causal operators

A neutral functional equation with causal operators has the form

$$(Vx)(t) = (Wx)(t), \quad t \in [t_0, T), \tag{1}$$

with both V and W causal operators on a given function space $E([t_0, T), \mathbb{R}^n)$. The functional differential equation counterpart has the form

$$\frac{d}{dt}(Vx)(t) = (Wx)(t), \quad t \in [t_0, T), \tag{2}$$

where V and W have the same meaning as in equation (1).

These types of functional equations are more general than those encountered in such basic references as J.K. Hale [1], and the literature mentioned therein.

It is possible to reduce neutral functional differential equations of the form (2) to those of form (1), by integration. Indeed, from (2) one derives

$$(Vx)(t) = c + \int_{t_0}^{t} (Wx)(s)\,ds, \tag{3}$$

where $c \in \mathbb{R}^n$ is a constant vector to be determined by means of supplementary conditions imposed on equation (3).

For instance, if the operator V has the property of fixed initial value, then c in (3) must be chosen equal to this value. We shall illustrate in this chapter several situations when the choice can be made adequately.

Returning now to equation (1), we notice that by denoting $U = V - W$, it can be written as $(Ux)(t) = \theta$. But the hypotheses we shall make, in order to derive existence results for equations (1) or (2), can be better formulated if we rely on the form (1), instead of $(Ux)(t) = \theta$. Actually, the last form suggests that we are dealing with a functional equation in implicit form. In our view, the theory of neutral functional or functional differential equations we shall present in this chapter constitutes an attempt to provide efficient results which do appear in the "normal" form, $x(t) = (Vx)(t)$.

The existing journal literature is mainly concerned with various types of functional or functional differential equations, particularly delay equations or integrodifferential equations, but also ordinary differential equations. For instance, a good deal of work has been done in connection with the "neutral" equation

$$x'(t) = f(t, x(t), x'(t)). \tag{4}$$

If one chooses the space $C^{(1)}([t_0, T), \mathbb{R}^n)$ as the underlying space, then (4) takes the form (1), with $(Vx)(t) = x'(t)$ and $(Wx)(t) = f(t, x(t), x'(t))$.

It is useful to point out the fact that associating with (4) an initial condition of the form $x(t_0) = x^0 \in \mathbb{R}^n$, the operator $(Vx)(t) = x'(t)$ has an inverse, namely

$$(V^{-1}y)(t) = x^0 + \int_{t_0}^t y(s)\,ds. \tag{5}$$

This feature will become a guideline in investigating the general equation (1).

We should like to emphasize the fact that by choosing conveniently the operators V and W in equation (1), as well as the underlying space $E([t_0, T), \mathbb{R}^n)$, one obtains a large variety of functional or functional differential equations. This statement will be (partially) illustrated in the next sections of this chapter.

Particular mention is due to N.V. Azbelev [2], [3] and his associates (N.V. Azbelev, V.P. Maksimov and L.F. Rakhmatullina [1], N.V. Azbelev and L.F. Rakhmatullina [2], N.V. Azbelev and P.M. Simonov [1]). See also the references under the names of L.M. Berezanskii, V.P. Maksimov, who have developed a theory of linear (and quasilinear) functional equations formally of the type

$$(Lx)(t) = f(t). \tag{6}$$

The quasilinear functional equation associated with (6) has the form

$$(Lx)(t) = (fx)(t), \tag{7}$$

in which f is, generally, a nonlinear operator.

It is obvious that equations (6) or (7) are special cases of equation (1).

As mentioned above, there exists a rich literature in regard to the theory of equations (6) and (7), also available in book form, and in English. This fact has motivated us to omit in our presentation this theory and refer the reader to the above-mentioned references.

Before we conclude this introductory section about neutral equations, let us make a few general remarks on equation (1).

If the operator V appearing in (1) has a continuous inverse on the underlying space, then equation (1) is equivalent to the following functional equation which has the "normal" form

$$x(t) = (V^{-1}(Wx))(t). \tag{8}$$

Equation (8) is exactly the type of functional equation investigated in Chapter 3. But, in order to be able to apply the existence result of Chapter 3, we need to have a causal operator in the right-hand side of (8). Since V^{-1} may exist but be noncausal, it follows that the causality of V^{-1} must be formulated as a hypothesis. Of course, in case V is an operator which is not causal, but has a causal inverse, the existence results of Chapter 3 can be used to obtain existence for the functional equation (1).

6.2 Existence results in the continuous case

Let us now formulate a general existence result for equation (1). The following hypotheses will provide the necessary background.

H_1 V and W are operators acting on the space $C([0, T], \mathbb{R}^n)$, such that V has a continuous causal inverse V^{-1}, while W is continuous, causal and compact.

H_2 The operator $(V^{-1}(Wx))(t)$ has the fixed initial value property.

Let us notice that the *fixed initial value property*, which we have defined in Chapter 2, plays an important role in the continuous case, but it is deprived of meaning when the underlying space consists of measurable functions.

Theorem 6.1 *Assume the operators V and W satisfy hypotheses H_1, H_2. Then, there exists a local solution of (1), in the space $C([0, \delta], \mathbb{R}^n)$, for some $\delta > 0$ ($\delta \leq T$).*

The proof of Theorem 6.1 follows immediately from the local existence theorem for functional equations of the form $x(t) = (Vx)(t)$ in the space of continuous functions (Theorem 3.1). Indeed, (1) is equivalent to $x(t) = (V^{-1}(Wx))(t)$, and the product $V^{-1}(W)$ is compact.

When applying Theorem 6.1, an important step is proving that V^{-1} does exist on the space C, and it is a causal operator.

We shall now discuss, in detail, the functional differential equation (2), in which the operator V is given by

$$(Vx)(t) = x(t) + g(x(\alpha(t))). \tag{9}$$

We shall reduce this equation to an equation of the form (1). More precisely, the functional differential equation is

$$\frac{d}{dt}[x(t) + g(x(\alpha(t)))] = (Wx)(t), \tag{10}$$

and we will associate to (10) the usual initial value condition $x(t_0) = x^0 \in \mathbb{R}^n$.

Concerning equation (10), we will make the following assumptions.

1   $g : C([0, T], \mathbb{R}^n) \to C([0, T], \mathbb{R}^n)$ is a contraction map on this space:

$$|g(x) - g(y)|_C \leq \lambda |x - y|_C,$$

where $\lambda \in [0, 1)$.

2   $W : C([0, T], \mathbb{R}^n) \to C([0, T], \mathbb{R}^n)$ is a continuous causal operator, taking bounded sets of C into bounded sets.

3   $\alpha : [0, T] \to [0, \infty)$ is continuous and such that $\alpha(0) = 0$ and $0 \leq \alpha(t) \leq t$, for any $t \in [0, T]$.

Let us point out a few elementary properties that will enable us to deal with a simpler functional equation of the form (1), equivalent to (10), and the initial value condition $x(0) = x^0$.

First, without loss of generality, we can assume $g(x^0) = \theta \in \mathbb{R}^n$. It is obvious that an additive constant in g does not make any difference. So, if $g(x)$ is replaced by $g(x) - g(x^0)$, in case $g(x^0) \neq \theta$, we can obtain the desired property.

Second, by the integration of both sides of (10) from 0 to t, we get

$$x(t) + g(x(\alpha(t))) = x^0 + \int_0^t (Wx)(s)\, ds. \tag{11}$$

One has to take into account $\alpha(0) = 0$ and $g(x^0) = \theta$.

Of course, if we differentiate both sides of (11), we will obtain (10). Finally, for $t = 0$ both sides in (11) reduce to x^0.

Therefore, we shall deal with the functional equation (11), under assumptions 1–3 formulated above.

In order to match the conditions of Theorem 6.1, we need to show that the operator V defined by (9) has an inverse operator on $C([0, T], \mathbb{R}^n)$, which is continuous and causal. This is, indeed, the case and we shall prove these properties. Let us notice that the equation $Vx = f$ is, in fact,

$$x(t) + g(x(\alpha(t))) = f(t), \quad f \in C([0, T], \mathbb{R}^n).$$

If we write this equation in the form

$$x(t) = -g(x(\alpha(t))) + f(t), \tag{12}$$

we see that the right-hand side is a contraction map on $C([0, T], \mathbb{R}^n)$. Hence, equation (12) has a unique solution $x \in C$, for every $f \in C$. This property says that for each $f \in C$ there exists $x = V^{-1} f$ uniquely determined, which proves the existence of V^{-1}. This operator is obviously causal, since to determine x from (12), on an interval $[0, \bar{t}]$, we need only the values of f for $t \leq \bar{t}$.

The continuity of the inverse V^{-1} follows from a recent result of T. A. Burton [3]. We refer the reader to this reference for details.

In conclusion, the operator V given by (9) has an inverse on the space $C([0, T], \mathbb{R}^n)$, which is both causal and continuous. These properties are required in hypothesis H_1. It is easy to check the validity of H_2, under our assumptions.

Therefore, based on the discussion carried out above, we can state the following result.

Corollary 6.2   *Consider the functional differential equation* (10), *under the initial condition* $x(0) = x^0 \in \mathbb{R}^n$. *Assume that the properties 1–3 formulated above are satisfied. Then, there exists a solution to equation* (10), *belonging to the space* $C([0, \delta], \mathbb{R}^n)$, $0 < \delta < T$, *for some* δ.

Remark 1   The compactness of the operator in the right-hand side of (11) is assured by the condition 2 on W. From

$$(Wx)(t) = x^0 + \int_0^t (Wx)(s)\, ds$$

one easily obtains the equicontinuity of the functions represented by the right-hand side of (11), when they belong to a bounded set in C.

Remark 2   The functional differential equation (10), obviously of neutral type and with causal operators, is also an equation with modified argument. We have in mind the term $g(x(\alpha(t)))$, in which $\alpha(t)$ satisfies the condition 3, formulated above. In Section 6.3 of this chapter, we will examine another case of modified argument, namely $g = g(x(t - h))$. An initial functional datum will be required.

Remark 3   Under the assumptions 1–3 the uniqueness of the solution in Theorem 6.1 does not necessarily hold. For instance, taking $g = \theta$ and $(Wx)(t) = f(t, x(t))$, the function f being continuous and such that uniqueness is not assured for the equation $\dot{x}(t) = f(t, x(t))$, we obtain an illustration of the nonuniqueness for equation (10).

Hence, it is necessary to add some extra conditions to secure the uniqueness of the solution of (2), under the initial condition $x(0) = x^0$.

It turns out that such a condition is the generalized Lipschitz condition

$$|(Wx)(t) - (Wy)(t)| \le \mu(t) \sup_{0 \le s \le t} |x(s) - y(s)|, \tag{13}$$

for all $x, y \in C$, and $\mu(t)$ a positive nondecreasing function on $[0, T]$.

The proof of the uniqueness can be conducted as follows.

If $x(t)$ and $y(t)$ are solutions of equation (10), satisfying the same initial condition $x(0) = y(0) = x^0$, then

$$x(t) - y(t) + g(x(\alpha(t))) - g(y(\alpha(t))) = \int_0^t [(Wx)(s) - (Wy)(s)] \, ds.$$

This leads, for $t \in [0, \delta]$, to

$$|x(t) - y(t)| - |g(x(\alpha(t))) - g(y(\alpha(t)))| = \int_0^t |(Wx)(s) - (Wy)(s)| \, ds.$$

Taking into account the fact that g is a contraction mapping on C, we can write

$$|x(t) - y(t)| - \lambda |x(\alpha(t))) - y(\alpha(t)))| \le \int_0^t \mu(s) \sup_{0 \le u \le s} |x(u) - y(u)| \, ds.$$

We can strengthen this inequality, and obtain

$$|x(t) \quad y(t)| \le \lambda \sup_{0 \le s \le t} |x(s) - y(s)| + \int_0^t \mu(s) \sup_{0 \le u \le s} |x(u) - y(u)| \, ds.$$

Since the right-hand side of the last inequality is nondecreasing in t, we can write for $z(t) = \sup_{0 \le s \le t} |x(s) - y(s)|$ the following inequality:

$$(1 - \lambda) z(t) \le \int_0^t \mu(s) z(s) \, ds, \quad t \in [0, \delta]. \tag{14}$$

For each $\varepsilon > 0$ (14) implies

$$z(t) < \varepsilon + (1 - \lambda)^{-1} \mu(\delta) \int_0^t z(s) \, ds, \quad t \in [0, \delta]. \tag{15}$$

The inequality (15) is of Gronwall type and one derives

$$z(t) \le \varepsilon \exp\{(1 - \lambda)^{-1} \mu(\delta) t\},$$

which leads to $z(t) \equiv 0$ on $[0, \delta]$. Therefore, condition (13), added to the conditions 1–3 formulated above, assures the uniqueness of the solution of equation (10), with initial condition $x(0) = x^0$. We shall now provide a global existence theorem for the functional differential equation (2), which is equivalent to the functional equation (3), with $c \in \mathbb{R}^n$ arbitrary.

In order to obtain a global result, we will impose some restrictions on the operator V^{-1} appearing in the equation (equivalent to (3))

$$x(t) = V^{-1}\left(c + \int_0^t (Wx)(s)\,ds\right),\tag{16}$$

while keeping the conditions on W. More precisely, the following result holds true.

Theorem 6.3 *Consider the functional differential equation (2), under the following assumptions: V is a homeomorphism between a bounded set of $C([0, T], \mathbb{R}^n)$ and the whole space $C([0, T], \mathbb{R}^n)$, while V^{-1} is a causal operator on $C([0, T], \mathbb{R}^n)$. Moreover assume that the operator W is continuous and causal on $C([0, T], \mathbb{R}^n)$, taking bounded sets into bounded sets.*

Then, there exists a solution of (2) in $C([0, T], \mathbb{R}^n)$. More precisely, equation (16) has a solution in $C([0, T], \mathbb{R}^n)$ for each $c \in \mathbb{R}^n$. Each solution of (16) also satisfies (3) and (2).

Proof We shall deal with equation (16), proving the existence of a solution in $C([0, T], \mathbb{R}^n)$. Let us notice that equation (16) has the "normal" form, which allows a straightforward application of fixed point arguments.

Under our assumptions, V^{-1} is causal and continuous on $C([0, T], \mathbb{R}^n)$. Also, the image of the space $C([0, T], \mathbb{R}^n)$, by the operator V^{-1}, is a bounded set in $C([0, T], \mathbb{R}^n)$. The operator

$$x(t) \to c + \int_0^t (Wx)(s)\,ds$$

is a continuous compact operator on $C([0, T], \mathbb{R}^n)$.

The compactness follows easily from the estimate

$$|(Wx)(t) - (Wx)(s)| \leq M|t - s|,$$

where $M > 0$ is an upper bound for the norm of all elements in WB, with B a bounded subset in $C([0, T], \mathbb{R}^n)$.

Consequently, the operator

$$x(t) \to V^{-1}\left(c + \int_0^t (Wx)(s)\,ds\right)\tag{17}$$

is continuous and compact in $C([0, T], \mathbb{R}^n)$, while the set $V^{-1}C([0, T], \mathbb{R}^n)$ is a bounded subset of $C([0, T], \mathbb{R}^n)$. In other words, the operator in (16) takes the whole space $C([0, T], \mathbb{R}^n)$ into a bounded set.

We can now apply Corollary 3.8 to equation (16), obtaining the (global) existence of a solution.

This ends the proof of Theorem 6.3.

Remark The condition of boundedness of the set $V^{-1}C([0, T], \mathbb{R}^n)$ in $C([0, T], \mathbb{R}^n)$ could be replaced by a growth condition on V^{-1}. See Theorem 3.7 which can serve as a model for obtaining results similar to Theorem 6.3.

6.3 Existence results in spaces of measurable functions

Equation (1), with the operator V given by

$$(Vx)(t) = x(t) + g(x(t - h)), \quad h > 0, \tag{18}$$

will be discussed now, taking $L^\infty_{\text{loc}}([t_0, T), \mathbb{R}^n)$ as the underlying space.

The operator V defined by (18) is obviously a delay type operator, and we will assume $h > 0$ to be a fixed number. Hence, we deal with a case of bounded delay. In this case, an initial functional condition must be assigned, of the form

$$x(t) = x_0(t) \in L^\infty([t_0 - h, t_0], \mathbb{R}^n). \tag{19}$$

By means of (19), we see from (18) that V is defined for any $x(t) \in L^\infty_{\text{loc}}([t_0, T), \mathbb{R}^n)$. In order to be sure that V takes its values in $L^\infty_{\text{loc}}([t_0, T), \mathbb{R}^n)$, we shall assume that $g : \mathbb{R}^n \to \mathbb{R}^n$ is a *homeomorphism* (onto). This condition will also serve to construct the inverse operator V^{-1} on the space $L^\infty_{\text{loc}}([t_0, T), \mathbb{R}^n)$.

To prove the existence of V^{-1} on $L^\infty_{\text{loc}}([t_0, T), \mathbb{R}^n)$, we need to solve the "functional" equation

$$x(t) + g(x(t - h)) = f(t), \quad t \in [0, T), \tag{20}$$

keeping in mind the initial condition (19). This can be achieved by the method of steps.

Indeed, let us determine $x(t) = x_1(t)$, on the interval $[t_0, t_0 + h]$, from the equation

$$x_1(t) + g(x_0(t - h)) = f(t).$$

This is (uniquely) possible and obviously $x_1(t) \in L^\infty([t_0, t_0 + h], \mathbb{R}^n)$.

By means of the recurrent formula

$$x_{k+1}(t) + g(x_k(t - h)) = f(t),$$

with validity on $[t_0 + (k - 1)h, t_0 + k]$, we determine $x_{k+1}(t)$, as soon as $x_k(t)$ is known. At each step, $x_k(t) \in L^\infty([t_0 + (k - 1)h, t_0 + kh], \mathbb{R}^n)$. Piecing together the $x_k's$, we obtain a function $x(t) \in L^\infty_{\text{loc}}([t_0, T), \mathbb{R}^n)$, which satisfies equation (20) almost everywhere on $[t_0, T]$.

We will now prove that the solution of equation (20) is unique. Indeed, if $x(t)$ and $y(t)$ are both solutions of (20) in $L^\infty_{\text{loc}}([t_0, T), \mathbb{R}^n)$, then we must have the relationship

$$x(t) - y(t) = g(y(t - h)) - g(x(t - h)), \tag{21}$$

valid on the interval $[t_0, T)$. Since $x(t - h)$ and $y(t - h)$ reduce to $x_0(t)$ on $[t_0, t_0 + h]$, it follows from (21) that $x(t) = y(t)$ on $[t_0, t_0 + h]$. Continuing by the method of steps, one concludes that $x(t) = y(t)$ on $[t_0, T)$.

The discussion carried out above shows that V^{-1} exists. Moreover, V^{-1} is a causal operator because $x(t) = (V^{-1}f)(t)$, as determined from (20), requires only the values of $f(t)$ for $t \leq \bar{t}$ in order to construct it on $[t_0, \bar{t}]$.

The continuity of V^{-1} can also be proven. The continuity of V^{-1} is equivalent to the fact that V takes any closed set of $L^\infty_{\text{loc}}([t_0, T), \mathbb{R}^n)$ into a closed set. In simpler terms, this means that if we choose a sequence of the form $\{Vx^{(m)}\}$, which converges in $L^\infty_{\text{loc}}([t_0, T), \mathbb{R}^n)$, then its limit is also of the form Vx, for some $x \in L^\infty_{\text{loc}}([t_0, T), \mathbb{R}^n)$.

Let us now assume that

$$Vx^{(m)} \to f \quad \text{as } m \to \infty,$$

in the space $L^\infty_{\text{loc}}([t_0, T), \mathbb{R}^n)$. Denote $f^{(m)}(t) = (Vx^{(m)})(t)$, $m \geq 1$. From (18) and (20) we obtain

$$f^{(m)}(t) = x^{(m)}(t) + g(x^{(m)}(t-h)) \to f(t), \tag{22}$$

in the space $L^\infty_{\text{loc}}([t_0, T), \mathbb{R}^n)$. Of course, each $x^{(m)}(t)$ satisfies the initial condition (19). Let us consider (22) on the initial $[t_0, t_0 + h]$. Then it becomes

$$f^{(m)}(t) + g(x_0(t-h)) \to f(t)$$

in the space $L^\infty([t_0, t_0 + h), \mathbb{R}^n)$. From (22) we derive that $\{f^{(m)}(t)\}$ and $\{x^{(m)}(t)\}$ are simultaneously convergent on $[t_0, t_0 + h]$, in the topology of the space $L^\infty([t_0, t_0 + h], \mathbb{R}^n)$. Now we consider (22) on the interval $[t_0 + h, t_0 + 2h]$ and by the same type of argument as above we obtain the convergence of the sequence $\{x^{(m)}(t)\}$ on the interval $(t_0 + h, t_0 + 2h]$. By repeating this argument we see that $\{x^{(m)}(t)\}$ converges in $L^\infty([t_0, t_1], \mathbb{R}^n)$, with arbitrary $t_1 < T$. This means exactly the convergence of $\{x^{(m)}(t)\}$ in $L^\infty_{\text{loc}}([t_0, T), \mathbb{R}^n)$.

We shall denote

$$x(t) = \lim_{m \to \infty} x^{(m)}(t), \quad \text{a.e. on } [t_0, T),$$

and letting $m \to \infty$ in (22), we obtain (20), which means $(Vx)(t) = f(t)$. This proves the continuity of the operator V^{-1} on $L^\infty_{\text{loc}}([t_0, T), \mathbb{R}^n)$. Implicitly, it has been shown that V is a homeomorphism of the space $L^\infty_{\text{loc}}([t_0, T), \mathbb{R}^n)$.

Theorem 6.4   *Consider the neutral functional equation (1), with V given by (18). Assume that g is a homeomorphism of the space $\mathbb{R}^n$ (onto), while W is causal, continuous and compact on $L^\infty_{\text{loc}}([t_0, T), \mathbb{R}^n)$.*

Then, there exists a solution of (1), belonging to some space $L^\infty([t_0, t_0 + \delta], \mathbb{R}^n)$, for $\delta > 0$ $(t_0 + \delta < T)$, satisfying the initial condition (19).

Proof   Writing equation (1) in the equivalent form

$$x(t) = (V^{-1}(Wx))(t),$$

and taking into account the fact that the operator in the right-hand side is continuous, compact and causal on $L^\infty_{\text{loc}}([t_0, T), \mathbb{R}^n)$, the existence of a solution (as a fixed point) follows by standard arguments.

For a global result of existence it is necessary to impose some growth condition on V^{-1}.

As an example of a functional equation of the form (1), with V given by (18), we will consider the following functional integral equation with delay

$$x(t) + g(x(t-h)) = \int_{t_0}^t k(t, s) x(s) \, ds + f(t), \tag{23}$$

together with the initial condition (19). If we keep the assumption that g is a homeomorphism of $\mathbb{R}^n$, it remains to assure the continuity and compactness of the integral operator

$$x(t) \to \int_{t_0}^t k(t, s) x(s) \, ds$$

on the space $L^\infty_{\mathrm{loc}}([t_0, T), \mathbb{R}^n)$. Since the matrix kernel $k(t, s)$ is defined on $\Delta = \{(t, s); \ t_0 \leq s \leq t < T\}$, a condition which guarantees the continuity is $|k| \in L^\infty_{\mathrm{loc}}(\Delta, \mathbb{R})$. If we strengthen somewhat this condition, assuming $|k| \in C_{\mathrm{loc}}(\Delta, \mathbb{R})$, then compactness in L^∞ is also assured, and Theorem 6.4 yields the existence of a local solution to (23).

6.4 More results on neutral functional differential equations

We shall be concerned in this section with functional differential equations with causal operators of the form

$$\frac{d}{dt}[x(t) + (Vx)(t)] = (Wx)(t). \tag{24}$$

Compared with the form (2), we notice that $(Vx)(t)$ has been replaced by $x(t) + (Vx)(t)$. The reason is that we can more easily formulate conditions for the invertibility of the operator

$$(V_1 x)(t) = x(t) + (Vx)(t), \tag{25}$$

in various function spaces, a feature allowing us to apply the method of reduction of neutral functional equations to the "normal" form $x(t) = (Tx)(t)$.

For instance, it is well known from the theory of linear operators that $I + V$, with $|V|$ small, is invertible (in a Banach space framework). There are other opportunities, especially when we deal with Fredholm theory for linear operators. More examples will be discussed in this section.

Let us start the investigation of equation (24), under the usual initial condition $x(0) = x^0 \in \mathbb{R}^n$.

The following basic assumptions will be made on the operators V and W:

1 The operators V and W are continuous causal operators on the space $C([0, T], \mathbb{R}^n)$.
2 V is also compact, and has the fixed initial value property.
3 The operator W takes bounded sets of $C([0, T], \mathbb{R}^n)$ into bounded sets.

Without loss of generality, we can assume that $(Vx)(0) = \theta$, for any $x \in C([0, T], \mathbb{R}^n)$. Indeed, since V has the fixed initial property, i.e., $(Vx)(0) = c \in \mathbb{R}^n$ for $x \in C$, it is obvious that the condition $(Vx)(0) = \theta$ can be satisfied if V is replaced by $V - cI$.

Integrating both sides of (24) from 0 to $t \in (0, T)$, we obtain the functional equation

$$x(t) + (Vx)(t) = x^0 + \int_0^t (Wx)(s)\, ds. \tag{26}$$

We have taken into account the initial condition $x(0) = x^0$, as well as $(Vx)(0) = \theta$.

The equivalence of (26) with (24) and the initial condition is almost obvious. Therefore, we shall concentrate on equation (26), and prove that, under conditions 1–3, there exists a solution to this equation.

Theorem 6.5 *Consider the neutral functional differential equation* (24), *under the initial condition* $x(0) = x^0$. *Assume that the operators V and W satisfy the hypotheses 1–3 formulated above.*

Then, there exists a positive number δ, $\delta \leq T$, such that (24) *has a solution in $C([0, \delta], \mathbb{R}^n)$, satisfying $x(0) = x^0$.*

Proof   Equation (26) is equivalent to

$$x(t) = -(Vx)(t) + \int_0^t (Wx)(s)\, ds + x^0, \tag{27}$$

which has the "normal" form $x(t) = (Tx)(t)$.

In order to derive the existence of a solution in the space $C([0, \delta], \mathbb{R}^n)$, with some $\delta > 0$ ($\delta \leq T$), we shall apply Theorem 3.1.

The continuity and causality of the operator in the right-hand side of (27) are obvious. It remains to check the compactness of the operator in the right side of (27). Since V is compact by our assumption 2, and also possesses the fixed initial value property, we have only to show that the operator

$$x(t) \to \int_0^t (Wx)(s)\, ds$$

is compact on $C([0, T], \mathbb{R}^n)$. This fact has been noticed in preceding sections (see, for instance, the proof of Theorem 6.2), when W takes bounded sets into bounded sets of $C([0, T], \mathbb{R}^n)$.

This ends the proof of Theorem 6.4.

Remark   If besides conditions 1–3, we assume the validity of the estimates

$$|x(t) + (Vx)(t)| \geq \gamma |x(t)|,$$

and

$$|(Wx)(t)| \leq \alpha |x(t)| + \beta,$$

for all $x \in C$, then from (26) we derive the integral inequality

$$\gamma |x(t)| \leq |x^0| + \beta t + \alpha \int_0^t |x(s)|\, ds. \tag{28}$$

This inequality is of Gronwall type and yields an estimate of the form

$$|x(t)| \leq M(|x^0|, \alpha, \beta, \gamma), \tag{29}$$

on the interval of existence of the solution $x(t)$. As we know, there is a solution defined on some interval $[0, \delta]$, $\delta > 0$. Based on the causality of the operators V and W, this solution can be continued (to the right). The estimate (29) shows that an "escape" time cannot exist. Therefore, one can conclude that a result of global existence is valid.

Let us consider now equation (24) in the spaces $L^p([0, T], \mathbb{R}^n)$, $1 \leq p < \infty$. We shall accept the following conditions on the operators V and W:

H_1 V and W are causal operators on the space $L^p([0, T], \mathbb{R}^n)$, $1 \leq p < \infty$.
H_2 V is continuous and compact, such that

$$\lim(Vx)(t) = \theta \quad \text{as } t \to 0+, \tag{30}$$

for each $x \in L^p([0, T], \mathbb{R}^n)$.
H_3 W is a continuous operator on $L^p([0, T], \mathbb{R}^n)$, taking bounded sets into bounded sets.

Before we state the existence result, let us point out that condition (30) is a generalization of the fixed initial property (used several times but in the case of the space C).

An example of a classical operator satisfying (30) is the Volterra integral operator

$$(Vx)(t) = \int_0^t k(t, s, x(s)) \, ds,$$

under adequate conditions on the kernel $k(t, s, x)$; for instance, assuming $|k(t, s, x(s))| \leq k_0(s)$ a.e. for $0 \leq s \leq t \leq T$ and $x \in L^p$, with k_0 integrable. Of course, k_0 can be dependent on x.

The initial condition $x(0) = x^0$ will be replaced by the condition

$$\lim x(t) = x^0 \quad \text{as } t \to 0+. \tag{31}$$

This fact is due to the circumstance that $x(t)$ may not be continuous. We shall see that $x(t) + (Vx)(t)$ is absolutely continuous, which does not imply absolute continuity for $x(t)$.

Theorem 6.6 *Consider the neutral functional differential equation* (24), *under initial condition* (31). *Assume that the operators V and W satisfy the conditions H_1, H_2 and H_3 formulated above.*

Then, there exists a solution $x(t)$ of this problem, defined on some interval $[0, \delta]$, $0 < \delta < T$, such that $x(t) \in L^p([0, \delta], \mathbb{R}^n)$, $1 \leq p < \infty$. Moreover, $x(t) + (Vx)(t)$ is an absolutely continuous function.

Proof First, we shall notice that the problem (24), (31) is equivalent in L^p to the functional equation

$$x(t) + (Vx)(t) = x^0 + \int_0^t (Wx)(s) \, ds. \tag{32}$$

Indeed, if $x(t) \in L^p$ is a solution of the problem, then $(Wx)(t) \in L^p \in L^1$, and both sides of (24) can be integrated on any interval $[0, t]$, $t \leq T$. If one takes into account H_2, the result of integration can be written in the form (32). Since the right-hand side of (32) is an absolutely continuous function, one can differentiate (32) almost everywhere, and obtain (24). Also, letting $t \to 0+$ in (32), one obtains (31).

Therefore, we have to prove the existence of an L^p-solution to the functional equation (32), i.e.,

$$x(t) = -(Vx)(t) + \int_0^t (Wx)(s) \, ds + x^0. \tag{33}$$

Equation (33) has the "normal" form $x(t) = (Tx)(t)$, and it remains to check the validity of the conditions assuring the existence of a solution in L^p. One can take as reference Theorem 3.3. The hypotheses H_1, H_2, H_3 provide for the causality, the continuity and the compactness of the operator in the right-hand side of the functional equation (33) (of course, on the space $L^p([0, T], \mathbb{R}^n)$, $1 \leq p < \infty$).

It is worthwhile mentioning that the operator

$$x(t) \to \int_0^t (Wx)(s) \, ds,$$

considered from L^p into C, is a compact operator (the Arzelà–Ascoli criterion is satisfied by H$_3$), which implies compactness from L^p into L^p.

At this point, we can directly apply Theorem 3.3, which ends the proof of Theorem 6.6.

Remark   Theorem 6.6 can be applied to neutral functional differential equations of the form

$$\frac{d}{dt}\left[x(t) + \int_0^t K(t, s, x(s))\,ds\right] = F(t, x(t)),$$

for instance. In this equation

$$(Vx)(t) = \int_0^t K(t, s, x(s))\,ds,$$

and

$$(Wx)(t) = F(t, x(t)).$$

We leave to the reader the task of formulating conditions on K and F, in such a way that the operators V and W satisfy hypotheses H$_1$, H$_2$ and H$_3$. Also, it is possible to make many other choices for the operator W in equation (24), including discrete or continuously distributed delays. See C. Corduneanu and M. Mahdavi [2], [3] for various special cases of the operators V and W.

In concluding this section, we notice that global existence is assured when V and W are linear operators. In the next section, this case will be discussed in some detail.

6.5 The linear and quasilinear cases

Let us consider now the neutral functional differential equation (24), or its equivalent (26), and assume that both V and W are linear/affine operators on the space $L^p_{\text{loc}}([0, T), \mathbb{R}^n)$, $1 \le p < \infty$. Then equation (26) can be rewritten in the standard form for linear equations

$$x(t) = (Lx)(t) + f(t), \quad t \in [0, T], \tag{34}$$

where

$$(Lx)(t) = \int_0^t [(Wx)(s) - (W\theta)(s)]\,ds - [(Vx)(t) - (V\theta)(t)] \tag{35}$$

and

$$f(t) = \int_0^t (W\theta)(s)\,ds - (V\theta)(t) + x^0. \tag{36}$$

We shall now formulate some variants of hypotheses H$_1$, H$_2$, H$_3$, which will guarantee global existence. These hypotheses are:

C$_1$ V and W are causal affine operators on $L^p_{\text{loc}}([0, T), \mathbb{R}^n)$, $1 \le p < \infty$.
C$_2$ V is compact on $L^p_{\text{loc}}([0, T), \mathbb{R}^n)$ and satisfies condition (30).

Theorem 6.7 *Consider the linear neutral functional differential equation* (24), *under initial condition* (31). *Assume the operators V and W satisfy the conditions* C_1 *and* C_2, *formulated above.*

Then, there exists a unique solution $x(t) \in L_{loc}^p([0, T), \mathbb{R}^n)$ *of the problem, such that* $x(t) + (Vx)(t)$ *is locally absolutely continuous.*

Proof The linearity and continuity of the operator L, given by (35), are obvious from its definition and hypotheses C_1, C_2. The compactness of V is assured by C_2. Let us point out that compactness in $L^p([0, T), \mathbb{R}^n)$ means compactness on each $L^p([0, a], \mathbb{R}^n)$, for each $a < T$. The compactness of the operator

$$x(t) \to \int_0^t [(Wx)(s) - (W\theta)(s)] \, ds,$$

which appears in the definition of $(Lx)(t)$, on the space $L_{loc}^p([0, T), \mathbb{R}^n)$, can be obtained by the same argument used in the proof of Theorem 6.6.

Consequently, one can apply Theorem 4.2 to equation (34), which ends the proof of Theorem 6.7.

Remark 1 A similar result is valid when the underlying space is the space of continuous functions $C([0, T), \mathbb{R}^n)$, in which convergence means uniform convergence on compact intervals in $[0, T)$. This case has been examined, for instance, in C. Corduneanu [10].

Remark 2 We shall also notice that the global character of the existence is not necessarily reduced to the linear case. The global results of Section 3.4 can be used, in the same way Theorem 4.2 has been used in the proof of Theorem 6.7, in order to derive global existence when linearity is replaced by Lipschitz type continuity.

In concluding this section, we shall examine the quasilinear case of equation (24), relying on Theorem 4.6 for the existence part.

To simplify somewhat the presentation, we shall assume that V is linear on $C([0, T], \mathbb{R}^n)$, while W is – in general – nonlinear.

The following hypotheses will be made on the operators V and W.

a V is causal, linear, continuous and compact on the space $C([0, T], \mathbb{R}^n)$.
b W is a causal continuous operator on the space $C([0, T], \mathbb{R}^n)$, taking bounded sets into bounded sets, and satisfying the growth condition

$$\limsup_{r \to \infty} [\phi(r)/r] = \gamma, \tag{37}$$

where

$$\phi(r) = \sup\{|(Wx)(t)|; \ |x(t)| \le r\},$$

with γ sufficiently small (to be made more precise below).

Theorem 6.8 *Consider the neutral functional differential equation* (24), *under condition* $x(0) = x^0 \in \mathbb{R}^n$. *Assume hypotheses* a, b *are satisfied by the operators V and W.*

Then, there exists a solution of this problem in $C([0, T], \mathbb{R}^n)$, such that $x(t) + (Vx)(t)$ is continuously differentiable.

Proof   We shall again use equation (32), which is equivalent to (24), with $x(0) = x^0$. By condition a, the operator $I + V$ is invertible on $C([0, T], \mathbb{R}^n)$. This is because the equation $x(t) + (Vx)(t) = f(t)$ is uniquely solvable for each $f \in C([0, T], \mathbb{R}^n)$, and the inverse $(I + V)^{-1} = U$ exists, is causal, and continuous.

With these observations, equation (32) can be written in the equivalent form

$$x(t) = (Ux^0)(t) + U\left(\int_0^t (Wx)(s)\, ds\right).$$
(38)

It is now obvious that (38) is a functional equation (nonneutral!), with causal operator in the right-hand side. Moreover, this operator is continuous and compact on $C([0, T], \mathbb{R}^n)$. The compactness is a consequence of the compactness of the operator

$$x(t) \rightarrow \int_0^t (Wx)(s)\, ds,$$

and of the continuity of $(I + V)^{-1} = U$.

Based on hypotheses b, we see that the conditions required by Theorem 4.6 are satisfied. One can easily find the estimate

$$\tilde{\phi}(r) = |U|_C T\phi(r)$$

which leads to the inequality

$$\limsup_{r \to \infty}[\tilde{\phi}(r)/r] \leq \gamma T|U|_C.$$
(39)

$\tilde{\phi}(r)$ has the same definition as $\phi(r)$, but for the operator $U(\int_0^t (Wx)(s)\, ds)$ instead of W.

Consequently, an upper estimate for r in (37) is $\gamma T|U|_C \leq 1$, or

$$\gamma < (T|U|_C)^{-1}.$$
(40)

Theorem 6.8 is thereby proven.

We suggest the reader apply Theorem 6.8 to the equation

$$\frac{d}{dt}\left[x(t) + \int_0^t k(t, s)x(s)\, ds\right] = A(t)x(t) + f(t),$$

formulating adequate conditions for existence.

7 Miscellanea (applications and generalizations)

7.1 A linear–quadratic optimal control problem with causal operators

The problem we shall discuss in this section is described by means of the linear differential system

$$\dot{x}(t) = (Ax)(t) + (Bu)(t), \quad t \in [0, T], \tag{1}$$

where $x : [0, T] \to \mathbb{R}^n$, $u : [0, T] \to \mathbb{R}^m$, while A and B denote causal linear operators on the space $L^2([0, T], \mathbb{R}^n)$, resp. $L^2([0, T], \mathbb{R}^m)$. The cost functional is chosen in the form

$$\mathcal{C}(x, u) = \int_0^T \{\langle (Px)(t), x(t) \rangle + \langle (Qu)(t), u(t) \rangle\} \, dt, \tag{2}$$

where P and Q stand for bounded self-adjoint operators on the space $L^2([0, T], \mathbb{R}^n)$, resp. $L^2([0, T], \mathbb{R}^m)$.

We will attach to (1) the initial condition

$$x(0) = \theta \in \mathbb{R}^n. \tag{3}$$

Condition (3) will determine a unique solution of (1) on $[0, T]$, as soon as u is assigned on that interval.

The optimal control problem consists in finding a control $\bar{u}$, within a certain class, such that

$$C(\bar{x}, \bar{u}) = \min C(x, u), \tag{4}$$

where $\bar{x}$ is the solution of (1), for $u = \bar{u}$, satisfying the initial condition (3). As is known, $\bar{u}$ will be the *optimal control* while $\bar{x}$ is the *optimal trajectory*.

The restriction on the control u is

$$u \in U \subset L^2([0, T], \mathbb{R}^m), \tag{5}$$

where U stands for a convex closed set.

The method we shall use to prove the existence (and uniqueness) of the optimal control $\bar{u}$ is elementary, and is based on changing adequately the scalar product/norm in the space $L^2([0, T], \mathbb{R}^m)$. We will obtain an equivalent norm and the conclusion will be an immediate consequence of the following property of Hilbert spaces: Any closed convex set in a Hilbert space contains a unique element of minimal norm.

First, let us notice that the relationship between u and x, as determined by (1) and (3), is of the form

$$x(t; u) = \int_0^t k(t, s)u(s)\, ds, \quad t \in [0, T],\tag{6}$$

with $k(t, s)$ of type n by m, uniquely defined by the operators A and B.

Indeed, if we denote by $X(t, s)$ the Cauchy matrix of the system $\dot{x}(t) = (Ax)(t)$, then (1) and (3) lead to

$$x(t; u) = \int_0^t X(t, s)(Bu)(s)\, ds.\tag{7}$$

For each fixed $t \in [0, T]$, the map $u \to x(t; u)$ is linear and continuous from $L^2([0, T], \mathbb{R}^m)$ into $\mathbb{R}^n$. According to the Riesz representation theorem, one can write

$$x(t; u) = \int_0^t k(t, s)u(s)\, ds, \quad t \in [0, T].\tag{8}$$

Since $x(t; u)$ is an absolutely continuous function, we may regard (8) as defining a map from $L^2([0, T], \mathbb{R}^m)$ into $AC([0, T], \mathbb{R}^n)$. Or, as a (continuous) map from $L^2([0, T], \mathbb{R}^m)$ into $\mathcal{C}([0, T], \mathbb{R}^n)$.

We know $k(t, s)$ to be a measurable kernel on $\Delta = \{(t, s); 0 \le s \le t \le T\}$, of type n by m. Based on the fact that the operator defined by (8) is acting continuously from $L^2([0, T], \mathbb{R}^m)$ into $\mathcal{C}([0, T], \mathbb{R}^n)$, we can write

$$\int_0^T dt \int_0^t |k(t, s)|^2\, ds < \infty\tag{9}$$

and

$$\lim_{h \to 0} \left\{ \int_0^t |k(t + h, s) - k(t, s)|^2 ds + \int_t^{t+h} |k(t, s)|^2 ds \right\} = 0.\tag{10}$$

For these Radon type properties, see, for instance C. Corduneanu [3] or P. Zabrejko *et al.* [1].

Second, let us define on $L^2([0, T], \mathbb{R}^m)$ another scalar product, by means of

$$\langle\langle u, v \rangle\rangle = \int_0^T \{\langle (Px)(t), y(t)\rangle + \langle (Qu)(t), v(t)\rangle\}\, dt,\tag{11}$$

where $x(t) = x(t; u)$, and $y(t) = x(t; v)$, for arbitrary $u, v \in L^2([0, T], \mathbb{R}^m)$. It is an elementary exercise to prove that $\langle\langle u, v \rangle\rangle$ defined by (11) satisfies the properties required for a scalar product. The only property requiring special attention is that related to the fact that $\langle\langle u, u \rangle\rangle = 0$ must imply $u(t) = \theta \in \mathbb{R}^m$, a.e. on $[0, T]$. This will be accomplished if, besides the self-adjointness of P and Q, we will assume $P \ge 0$ and $Q > 0$. Indeed, from $\langle\langle u, v \rangle\rangle = 0$ there follows $\langle (Qu)(t), u(t)\rangle = 0$ a.e. on $[0, T]$. Hence, $u(t) = 0$ a.e. on $[0, T]$ because $Q > 0$ means

$$\int_0^T \langle (Qu)(t), u(t)\rangle\, dt \ge \lambda \int_0^T |u(t)|^2\, dt\tag{12}$$

for some $\lambda > 0$, and any $u \in L^2([0, T], \mathbb{R}^m)$.

On the other hand, by (12) and $P \geq 0$ we can write

$$\lambda \int_0^T |u(t)|^2 \, dt \leq \int_0^T \{\langle (Px)(t), x(t) \rangle + \langle (Qu)(t), u(t) \rangle \} \, dt \tag{13}$$

for any $u \in L^2([0, T], \mathbb{R}^m)$. Inequality (13) means $x|u|_2 \leq C(x, u)$ for any $u \in L^2([0, T], \mathbb{R}^m)$, with $x = x(t; u)$.

Therefore, it remains to show that

$$C(x, u) \leq \Lambda \int_0^T |u(t)|^2 \, dt \tag{14}$$

for some $\Lambda > 0$, to conclude the equivalence of the norms induced on $L^2([0, T], \mathbb{R}^m)$ by the scalar products $\langle \cdot, \cdot \rangle$ and $\langle\langle \cdot, \cdot \rangle\rangle$. This fact is necessary in order to derive the closedness of U in the new topology, induced by the norm $\langle\langle v, u \rangle\rangle^{1/2}$.

The inequality (14) will follow from several inequalities we shall establish. Due to the boundedness of the operator P on the space $L^2([0, T], \mathbb{R}^m)$, we can write

$$\int_0^T \langle (Px)(t), x(t) \rangle \, dt \leq C_1 \int_0^T |x(t)|^2 \, dt, \tag{15}$$

for some $C_1 > 0$.

On the other hand, from (8) and (9) we derive

$$\int_0^T |x(t)|^2 \, dt \leq C_2 \int_0^T |u(t)|^2 \, dt, \tag{16}$$

where $x = x(t; u)$ and $u \in L^2([0, T], \mathbb{R}^m)$, for some $C_2 > 0$.

Since

$$\int_0^T \langle (Qu)(t), u(t) \rangle \, dt \leq C_3 \int_0^T |u(t)|^2 \, dt, \tag{17}$$

because of the boundedness of Q, with $C_3 > 0$, we obtain from (15)–(17) the inequality (14), where $\Lambda = C_1 C_2 + C_3$.

The inequalities (12) and (14) lead to the needed inequality for the equivalence of the norms, namely

$$\lambda \int_0^T |u(t)|^2 \, dt \leq C(x, u) \leq \Lambda \int_0^T |u(t)|^2 \, dt, \tag{18}$$

with $x = x(t; u)$, and arbitrary $u \in L^2([0, T], \mathbb{R}^m)$.

From (11) it is obvious that

$$\langle\langle u, u \rangle\rangle = C(x, u), \tag{19}$$

for any $u \in L^2([0, T], \mathbb{R}^m)$, which allows us to rewrite (18) in the form

$$\lambda |u|_2^2 \leq \langle\langle u, u \rangle\rangle \leq \Lambda |u|_2^2. \tag{20}$$

As pointed out in Chapter 2, the inequality (20) is necessary and sufficient for the equivalence of the norms induced by the scalar products $\langle \cdot, \cdot \rangle$ and $\langle\langle \cdot, \cdot \rangle\rangle$.

But (19) tells us that $\mathcal{C}(x, u)$ attains the minimum only when $\langle\langle u, u \rangle\rangle$ is minimal.

At this point we can apply the above quoted result about the existence and uniqueness of the element of minimum norm in a closed convex subset of a Hilbert space.

To summarize the discussion carried out above, we can formulate the following result.

Theorem 7.1   *Consider the linear quadratic optimal control problem defined by (1)–(4), and assume the following hypotheses are satisfied:*

1   *The operators A and B are linear, continuous and causal on the spaces $L^2([0, T], \mathbb{R}^n)$, resp. $L^2([0, T], \mathbb{R}^m)$.*
2   *The linear operators P and Q, defined on the spaces $L^2([0, T], \mathbb{R}^n)$ resp. $L^2([0, T], \mathbb{R}^m)$ are self-adjoint and bounded on these spaces.*
3   *P is nonnegative, while Q is strictly positive.*
4   *The control set $U \subset L^2([0, T], \mathbb{R}^m)$ is closed and convex.*

Then, there exists a unique control $\bar{u} \in U$, such that (4) is satisfied.

Remark   The fact that the initial condition (3) is of a particular form does not restrict the generality of the result. Indeed, if we replace (3) by the more general conditions $x(0) = x^0 \in \mathbb{R}^n$, and denote by $x_0(t)$ the corresponding solution of (1), then letting $x = y + x_0(t)$ in (1) one obtains $\dot{y}(t) = (Ay)(t) + (Bu)(t)$, while $y(0) = \theta$.

Before concluding this section, let us sketch an approximation procedure for the optimal control.

We shall start from the fact that, under our assumptions, the input–output relationship is given by (8). In other words, we disregard the fact that (8) is a consequence of (1), (3).

For each $U \subset L^2([0, T], \mathbb{R}^m)$, $x(t, u)$ given by (8) is in $AC([0, T], \mathbb{R}^n) \subset L^2([0, T], \mathbb{R}^n)$. We want to find (by approximation) the control $\bar{u} \in U$, such that (4) holds, the minimum being taken with respect to $u \in U$.

Let us choose an orthonormal basis $\{\varphi_k\}$ in $L^2([0, T], \mathbb{R}^m)$, and for each $N > 0$ consider the controls of the form

$$u = \sum_{k=1}^{N} C_k \varphi_k, \tag{21}$$

where C_k are arbitrary real numbers, subject only to the restriction $u \in U$. From (21) we obtain

$$Qu = \sum_{k=1}^{N} C_k \psi_k, \tag{22}$$

where

$$\psi_k(t) = (Q\varphi_k)(t), \quad k = 1, 2, \dots, N, \tag{23}$$

and also

$$x(t) = x(t; u) = \sum_{k=1}^{N} C_k \chi_k(t), \tag{24}$$

with

$$\chi_k(t) = \int_0^t k(t, s)\varphi_k(s)\,ds. \tag{25}$$

Let us further denote

$$\mu_k(t) = (P\chi_k)(t), \quad k = 1, 2, \ldots, N, \tag{26}$$

and notice that

$$\mathcal{C}(x, u) = \sum_{k=1}^N \sum_{j=1}^N \gamma_{kj} C_k C_j, \tag{27}$$

where

$$\gamma_{kj} = \int_0^T \{\langle \mu_k(t), \chi_j(t) \rangle + \langle \psi_k(t), \varphi_j(t) \rangle\}\,dt, \tag{28}$$

while $\gamma_{kj} = \gamma_{jk}$, due to the self-adjointness of P and Q.

Now, from (27), we see that $\mathcal{C}(x, u)$ is a quadratic form in C_k, $k = 1, 2, \ldots, N$, with C_k such that u given by (21) satisfies $u \in U$. This implies a constraint on the coefficient, $C_k, k = 1, 2, \ldots, N$.

Finding $\min \mathcal{C}(x, u)$, with $\mathcal{C}(x, u)$ given by (27) and under the constraint $u \in U$, means solving an extremal problem in finite dimension ($\mathbb{R}^N$), which leads to the optimal control in the class defined by (21), namely

$$\bar{u}_N = \sum_{k=1}^N \overline{C}_k \varphi_k, \tag{29}$$

It is well known from the theory of variational methods that the sequence $\{\mathcal{C}(\bar{x}_N, \bar{u}_N)\}$ is nondecreasing and converges to $\min \mathcal{C}(x, u)$, $u \in U$.

As an illustration of the constraint imposed by $u \in U$, when u is given by (21), we will choose U to be the ball of radius $R > 0$, centered at some $u_0 \in L^2([0, T], \mathbb{R}^m)$. Then, the condition for u, given by (21), to belong to this ball takes the form

$$\sum_{k=1}^N C_k^2 - 2\sum_{k=1}^N a_k C_k + |u_0|_2^2 - R^2 \le 0, \tag{30}$$

where

$$a_k = \int_0^T \langle u_0(t), \varphi_k(t) \rangle\,dt, \quad k = 1, 2, \ldots, N.$$

Therefore, in this particular case for U, the problem of finding $\overline{C}_k$ which appear in (29) is equivalent to minimizing the quadratic form (27), under the constraint (30).

In the very special case $u_0 = \theta$, the inequality (30) becomes

$$\sum_{k=1}^N C_k^2 \le R^2,$$

which can be easily handled.

7.2 A maximum principle approach

We shall now consider an optimal control problem which can be described by means of the functional differential system

$$\dot{x}(t) = (Lx)(t) + f(t, x(t); u(t)), \quad t \in [0, T], \tag{31}$$

where L stands for a linear continuous operator, of causal type, on the space $L^2([0, T], \mathbb{R}^n)$, while $f(t, x; u)$ is a map from $[0, T] \times \mathbb{R}^n \times U, U \subset \mathbb{R}^m$, into $\mathbb{R}^n$. The set U provides the constraint for the control variable $u : u(t) \in U, t \in [0, T]$.

As we can see from (31), in the right-hand side we have a causal operator, which has an operator component $(Lx)(t)$, as well as a Niemytskii type operator $f(t, x(t); u(t))$. This mixed type of term will enable us to reduce the problem to the case when the input–output equation is a classical Volterra integral equation. Then, a result provided in C. Corduneanu [10] will lead to the formulation of the maximum principle related to (31) under certain restrictions for the control u, with a cost functional of the form

$$\mathcal{C}(x, u) = g(x(T)). \tag{32}$$

Using the integral representation for the solution of the linear equation $\dot{x}(t) = (Lx)(t) + f(t)$, with initial condition

$$x(0) = x^0 \in \mathbb{R}^n, \tag{33}$$

we find that the control u and the trajectory $x = x(t, u)$ must be related by the integral equation

$$x(t) = X(t, 0)x^0 + \int_0^t X(t, s) f(s, x(s); u(s)) \, ds, \tag{34}$$

where $X(t, s)$ denotes the Cauchy matrix associated with the corresponding linear system.

As seen from (34), the presence of the causal operator, in its abstract form, has been replaced by the classical Volterra operator appearing in the right-hand side of (34).

We shall now list the terminal constraints of our problem, namely

$$g_i(x(T)) \leq 0, \quad i = 1, 2, \ldots, p, \tag{35}$$

$$g_i(x(T)) = 0, \quad i = p + 1, \ldots, q. \tag{36}$$

Concerning the set U, which provides the constraints for the control u, we will assume that U is closed.

Remark   The choice of $\mathcal{C}(x, u)$ in the form (32) may appear as a very particular case. Usually, one would expect a cost functional of the form

$$\mathcal{C}(x, u) = \int_0^T f^0(t, x(t); u(t)) \, dt.$$

As it is well known from the theory of optimal processes, if one introduces one more (scalar) variable in the system by the formula

$$y(t) = \int_0^t f^0(s, x(s); u(s)) \, ds, \quad t \in [0, T],$$

then $\mathcal{C}(x, u) = y(T)$, i.e., a functional of the form (32). Of course, this procedure requires some conditions on f^0, but it is generally agreed that a cost functional of the form (32) provides enough generality to the optimal control problem.

To summarize the above discussion, we formulate our problem in the following terms.

Find a control $u \in U, u : [0, T] \to \mathbb{R}^m$, such that $x(t)$ defined by (31) and (33) minimizes the cost functional (32), under the constraints (35) and (36).

The following result, based on the *maximum principle* in optimal control theory, provides necessary conditions for the couple u, x.

Theorem 7.2 *Consider the optimal control problem described by equations (31)–(33), and the constraints (35) and (36). Assume the following conditions are satisfied:*

1   *L is a linear, continuous and causal operator on the space $L^2([0, T], \mathbb{R}^n)$.*
2   *$f : [0, T] \times \mathbb{R}^n \times U \to \mathbb{R}^n$ is continuous, together with its Jacobian matrix $(\partial f / \partial x)$, with $U \subset \mathbb{R}^m$ a closed set.*
3   *f and $(\partial f / \partial x)$ are locally bounded on $[0, T] \times \mathbb{R}^n \times U$, i.e., there exist two positive numbers $M(K)$ and $N(K)$, for each compact set $K \subset \mathbb{R}^n$, such that*

$$|f(t, x; u)| \le M(K), \quad |\partial f / \partial x| \le N(K), \tag{37}$$

   for $(t, x; u) \in [0, T] \times K \times U$.
4   *$g, g_i : \mathbb{R}^n \to \mathbb{R}, \quad i = 1, 2, \ldots, q$, are continuously differentiable maps.*

Then, among all admissible measurable controls $u : [0, T] \to U$, there exists an optimal control $\bar{u}$, such that $g(\bar{x}(T))$ is minimal and all constraints (35), (36) are satisfied by $\bar{x}(t)$ – the trajectory of (31) satisfying (33).

Moreover, the optimal pair $(\bar{x}, \bar{u})$ must satisfy the following relationships:

i   There exist constant multipliers λ and $\lambda_j, j = 1, 2, \ldots, q$, and a map $\psi : [0, T] \to \mathbb{R}^n$, such that

$$|\lambda| + \sum_{j=1}^{q} |\lambda_j| = 1, \lambda_j \le 0, \quad j = 1, 2, \ldots, p, \tag{38}$$

and

$$\lambda_j g_j(\bar{x}(T)) = 0, \quad j = p+1, p+2, \ldots, q, \tag{39}$$

while

$$\psi(t) = \left[-dX(T, t) + \int_t^T \psi(s) X(s, t) \, ds \right] \tag{40}$$

with

$$-d = \lambda \frac{\partial}{\partial x} g(\bar{x}(T)) + \sum_{j=1}^{q} \lambda_j \frac{\partial}{\partial x} g_j(\bar{x}(T)). \tag{41}$$

ii   For almost all $t \in [0, T]$, one has

$$\left[dX(T, t) + \int_t^T \psi(s) X(s, t) \, ds \right] \cdot f(t, \bar{x}(t); \bar{u}(t))$$

$$= \max_{v \in U} \left\{ \left[dX(T, t) + \int_t^T \psi(s) X(s, t) \, ds \right] \cdot f(t, \bar{x}(t); v) \right\}. \tag{42}$$

As mentioned above, the proof of Theorem 7.2 follows from Theorem 6.2.6 in C. Corduneanu [10], in which one must take $k(t, x, u) = \theta$ (see this reference for more details about this kind of problem, in the framework of classical Volterra integral equations).

Remark The continuity of $X(t, s)$, which is required in Theorem 6.2.6 in our book [10], follows by elementary estimates, if one takes into account that

$$X(t, s) = I + \int_s^t \tilde{k}(t, s) \, ds,$$

where $\tilde{k}(t, s)$ is the resolvent kernel corresponding to the kernel $k(t, s)$ from the representation

$$\int_0^t (Lx)(s) \, ds = \int_s^t k(t, s) x(s) \, ds,$$

which has been discussed in Chapter 4.

7.3 Asymptotic behavior in second-order systems

These results generalize some results in C. Corduneanu [10], and have been obtained jointly with M. Mahdavi [1].

So far, we have not considered systems of equations of higher order, with causal operators. In this section we will investigate systems of the form

$$\ddot{x}(t) + (L\dot{x})(t) + \operatorname{grad} F(x(t)) = f(t), \tag{43}$$

in which L stands for a linear causal operator acting on the space $L^p(\mathbb{R}_+, \mathbb{R}^n)$, $1 < p < \infty$, $F : \mathbb{R}^n \to \mathbb{R}$ is a differentiable map, while $f \in L^p(\mathbb{R}_+, \mathbb{R}^n)$.

The dynamical interpretation of equation (43) is as follows: obviously, it can be rewritten in the form $m \overrightarrow{a} = \overrightarrow{F}$, in which the total force $\overrightarrow{F}$ is the result of action of the three distinct forces; first, $-(L\dot{x})(t)$ is a resistance or viscosity force; second, $\operatorname{grad} F(x(t))$ represents a conservative field; third, $f(t)$ can be regarded as noise, which tends to disappear as $t \to \infty$.

We shall associate to (43) the usual initial conditions (position and velocity)

$$x(0) = x^0, \quad \dot{x}(0) = v^0, \tag{44}$$

where $x^0, v^0 \in \mathbb{R}^n$.

As seen in Chapter 2, it is relatively simple to formulate conditions under which equation (43) has a local solution satisfying (44). But our aim is to obtain global existence for the nonlinear equation (43), on the whole semi-axis $\mathbb{R}_+ = [0, \infty)$, and provide some information about the asymptotic behavior of its solutions. To be more specific, we shall be interested in obtaining existence of bounded solutions on $\mathbb{R}_+$.

We shall make the following assumptions on the operator L, the nonlinear potential $F(x)$, and the term $f(t)$.

H_1 The operator L is a linear causal operator, continuously acting on the space $L^p(\mathbb{R}_+, \mathbb{R}^n)$, $1 < p < \infty$, such that the following "positiveness" condition is satisfied for all $t \in \mathbb{R}_+$:

$$\int_0^t \langle (Ly)(s), y(s) \rangle ds \geq \delta \left(\int_0^t |y(s)|^q ds \right)^{2/q}, \tag{45}$$

where $\delta > 0$ is a constant, q is the conjugate index to p ($p^{-1}+q^{-1}=1$), and $y \in L^p \cap L^q$.
   We notice that in case $p = 2$, condition (45) becomes the usual condition of positiveness for Hilbert spaces.

H_2 The map $F : \mathbb{R}^n \to \mathbb{R}^n$ is differentiable and satisfies the coerciveness condition

$$F(x) \to \infty \quad \text{and} \quad |x| \to \infty. \tag{46}$$

H_3 The map $f : \mathbb{R}_+ \to \mathbb{R}^n$ is such that

$$f \in L^p(\mathbb{R}_+, \mathbb{R}^n), \quad 1 < p < \infty. \tag{47}$$

Under these three hypotheses we can proceed to investigate the existence and behavior of solutions of (43), (44), on the semi-axis $\mathbb{R}_+$.

Theorem 7.3   *Consider equation (43), under initial conditions (44). Assume the hypotheses H_1, H_2 and H_3 are satisfied. Then any solution of (43), with arbitrary x^0, $v^0 \in \mathbb{R}^n$, is defined on $\mathbb{R}_+$ and remains bounded on $\mathbb{R}_+$, together with its first derivative.*

Proof   Let us notice that equation (43) is equivalent to the first-order system

$$\dot{x}(t) = v(t), \quad \dot{v}(t) + (Lv)(t) + \operatorname{grad} F(x(t)) = f(t), \tag{48}$$

which is a functional differential system with causal operators. If one integrates both sides of each equation in (48), and takes into account the initial conditions (44), one obtains the system

$$x(t) = x^0 + \int_0^t v(s)\, ds,$$

$$v(t) = v^0 + \int_0^t f(s)\, ds + \int_0^t (Lv)(s)\, ds + \int_0^t \operatorname{grad} F(x(s))\, ds,$$

which is of the form $X(t) = (VX)(t)$, where V stands for a causal operator, continuous and compact on the space $L^p(\mathbb{R}_+, \mathbb{R}^{2n})$. This remark tells us that a local existence theorem is valid for (43). If $x(t)$ and $v(t)$ exist on some interval $[0, T]$, $T < \infty$, then it is possible to extend them on a larger interval, say $[0, T+\delta]$, $\delta > 0$ (without uniqueness). If $x(t)$ and $v(t)$ are defined on a semi-open interval $[0, T)$, $T < \infty$, then the extension (continuation) of this solution to a larger interval will be possible only in case $\lim x(t)$ and $\lim v(t)$ exist as $t \uparrow T$. Of course, we mean finite limits.

Let us now consider a solution $x(t)$ of equation (42), satisfying (44), which does exist on some interval $[0, T)$. One obtains by scalar multiplication by $\dot{x}(t)$

$$\frac{1}{2}\frac{d}{dt}|\dot{x}(t)|^2 + \langle (L\dot{x})(t), \dot{x}(t)\rangle + \frac{d}{dt}F(x(t)) = \langle f(t), \dot{x}(t)\rangle. \tag{49}$$

Integrating both sides of (49) from 0 to t, $0 < t < T$, one obtains the relationship

$$\frac{1}{2}|\dot{x}(t)|^2 + \int_0^t \langle (L\dot{x})(s), \dot{x}(s)\rangle ds + F(x(t)) = \frac{1}{2}|v^0|^2 + F(x^0) + \int_0^t \langle f(s), \dot{x}(s)\rangle ds. \tag{50}$$

Taking condition (45) into account, and applying the Cauchy–Schwarz and Hölder inequalities in the right-hand side of (50), we obtain the following inequality:

$$\frac{1}{2}|\dot{x}(t)|^2 + \delta \left(\int_0^t |\dot{x}(s)|^q \, ds\right)^{2/q} + F(x(t))$$
$$\leq \frac{1}{2}|v^0|^2 + F(x^0) + \frac{1}{2\varepsilon}\left(\int_0^t |\dot{x}(s)|^p \, ds\right)^{2/p} + \frac{\varepsilon}{2}\left(\int_0^t |\dot{x}(s)|^q \, ds\right)^{2/q}. \tag{51}$$

For $\varepsilon < 2\delta$, inequality (51) leads to

$$\frac{1}{2}|\dot{x}(t)|^2 + \left(\delta - \frac{\varepsilon}{2}\right)\left(\int_0^t |\dot{x}(s)|^q ds\right)^{2/q} + F(x(t))$$
$$\leq \frac{1}{2}|v^0|^2 + F(x^0) + \frac{1}{2\varepsilon}\left(\int_0^\infty |f(s)|^p \, ds\right)^{2/p}. \tag{52}$$

But the right-hand side of (52) is a constant (in t), say $\mathcal{C}(x^0, v^0, f)$, while the middle term in the left-hand side is positive for $t \in [0, T)$. Hence, we can write

$$\frac{1}{2}|\dot{x}(t)|^2 + F(x(t)) \leq \mathcal{C}(x^0, v^0, f), \tag{53}$$

which is valid for any $t \in [0, T)$. Since $\dot{x}(t)$ remains bounded on $[0, T)$, it means (Cauchy's criterion) that $\lim x(t)$ exists as $t \uparrow T$, and is finite. The same conclusion is valid for $v(t)$, as one can easily derive from the second equation (46). Therefore, both $x(t)$ and $v(t)$ have finite limits as $t \uparrow T$, which means that $x(t)$ can be extended beyond T.

The above established property shows that $x(t)$ and $\dot{x}(t)$ cannot have a finite escape time. Therefore, all solutions of (42), continued at the right until becoming saturated, must be defined on $\mathbb{R}_+$.

If one now looks to (52), which is now valid on $\mathbb{R}_+$, one sees that

$$\dot{x}(t) \in L^q(\mathbb{R}_+, \mathbb{R}^n). \tag{54}$$

Also, from (53) one derives that both $x(t)$ and $\dot{x}(t)$ must be bounded on $\mathbb{R}_+$.
This ends the proof of Theorem 7.3.

Remark 1 From (54) and the boundedness of $\dot{x}(t)$ on $\mathbb{R}_+$ one obtains the more accurate inclusion

$$\dot{x}(t) \in L^r(\mathbb{R}_+, \mathbb{R}^n), \quad q \le r \le \infty. \tag{55}$$

This shows that the velocity of the dynamical system described by (43) cannot be sustained as t increases. In other words, the dynamics of the system has the tendency to be "quiet" when t becomes large.

Remark 2 If instead of equation (43) one considers the special case

$$\ddot{x}(t) + (L\dot{x})(t) + F(x(t)) = \theta, \tag{56}$$

and the positiveness condition (45) is replaced by the weaker condition

$$\int_0^t \langle (Ly)(s), y(s) \rangle ds \ge 0, \tag{57}$$

for $t \in \mathbb{R}_+$ and $y(t) \in L^p \cap L^q$, then the conclusion of Theorem 7.3 remains valid.
Indeed, letting $f(t) = \theta$ in (49), and taking into account (57), one obtains

$$\tfrac{1}{2}|\dot{x}(t)|^2 + F(x(t)) \le \tfrac{1}{2}|v^0|^2 + F(x^0), \tag{58}$$

which can be processed in the same way as inequality (51).

Remark 3 Under extra conditions on the operator L, it is possible that (43) has a "limiting equation" which turns out to be an ordinary differential equation, namely

$$\ddot{z}(t) + \operatorname{grad} F(z(t)) = 0.$$

For some results of this nature, but in the case of first-order equations, see C. Corduneanu [10].

Remark 4 If equation (43) is replaced by the equation

$$\ddot{x}(t) + A|x(t)|^\alpha \dot{x}(t) + (L\dot{x})(t) + \operatorname{grad} F(x(t)) = f(t),$$

with $A, \alpha > 0$, then under condition (57) for L, one can prove a result similar to that of Theorem 7.3. Actually, one can show that

$$\lim |\dot{x}(t)| = 0 \quad \text{as } t \to \infty,$$

which strengthens the above observation about the "quiet" behavior of the system, for large values of t.

The details of the proof of this result can be found in C. Corduneanu and M. Mahdavi [1].

7.4 Global existence for the equation $x(t) = (Lx)(t) + (Nx)(t)$

We shall now investigate the existence of a global solution to the equation

$$x(t) = (Lx)(t) + (Nx)(t), \tag{59}$$

with L a linear operator and N a nonlinear operator, both acting on the space $L^p([0, T], \mathbb{R}^n)$, $1 \le p < \infty$. The method of proof will be based on an alternative of the fixed point principle of Leray–Schauder type (see for instance, D. O'Regan and R. Precup [1]).

Let us first formulate this principle, which we shall use below.

Let E be a real Banach space and $K \subset E$ a convex set. Assume $U \subset K$ is open, and $\theta \in U$. If $T : \overline{U} \to K$ is a continuous and compact map, then either T has a fixed point in $\overline{U}$, or there exist $y \in \partial U$ and $\lambda \in (0, 1)$, such that $y = \lambda T y$.

In regard to equation (59), we shall make the following assumptions:

1   The linear operator $L : L^p([0, T], \mathbb{R}^n) \to L^q([0, T], \mathbb{R}^n)$, $1 \le p < \infty$, is continuous, causal and compact.
2   The operator $N : L^p([0, T], \mathbb{R}^n) \to L^p([0, T], \mathbb{R}^n)$ is continuous and compact.
3   There exists $M_0 > 0$, independent of $\lambda \in (0, 1)$, such that

$$|x|_p = \left(\int_0^T |x(s)|^p ds \right)^{1/p} \ne M_0, \tag{60}$$

for any solution (if any) x of the equation

$$x(t) = (Lx)(t) + \lambda(Nx)(t), \quad \lambda \in (0, 1). \tag{61}$$

Under the above formulated conditions on L and N, the following existence result holds true.

Theorem 7.4   *Consider equation (59), and assume conditions 1–3 above are satisfied. Then, there exists a solution $x(t) \in L^p([0, T], \mathbb{R}^n)$ of the equation (59).*

Remark   It is obvious that under conditions 1 and 2 above, equation (59) has a local solution on some interval $[0, \delta]$, $\delta \le T$, when N is also a causal operator. This is guaranteed by the fact that the right-hand side of (59) is continuous and compact on $L^p([0, T], \mathbb{R}^n)$. Theorem 3.3 applies immediately.

Proof of Theorem 7.4   Let us consider first the auxiliary (linear) equation related to (59)

$$x(t) = (Lx)(t) + g(t), \tag{62}$$

in the space $L^p([0, T], \mathbb{R}^n)$, $1 \le p < \infty$.

As we know from Chapter 4, the unique solution of (62) can be expressed by means of the resolvent operator $R = (I - L)^{-1}L$ in the form

$$x(t) = g(t) + (Rg)(t). \tag{63}$$

This formula allows us to rewrite equation (63) in the form

$$x(t) = (Nx)(t) + (RN)(t). \tag{64}$$

Similarly, the parametrized equation (61) can be rewritten as

$$x(t) = \lambda[(Nx)(t) + (RNx)(t)]. \tag{65}$$

Let us now choose the set U, which appears in the statement of the Leray–Schauder principle, in the form

$$U = \{u; \ u \in L^p([0, T], \mathbb{R}^n), \ |u|_p < M_0\}. \tag{66}$$

We also denote $E = K = L^p(0, T], \mathbb{R}^n)$. Instead of dealing directly with the parametrized equation (61), we shall deal with the equivalent equation (65), which has the form $y = \lambda Ty, \lambda \in (0, 1)$, with $T = N + RN$.

Since R is (like L) continuous and compact, there follows continuity and compactness of the operator $T = N + RN$, on the space $L^p([0, T], \mathbb{R}^n) = E$.

Using condition 3 we see that equation (65) cannot have a solution in ∂U (i.e., such that $|x|_p = M_0$, for any $\lambda \in (0, 1)$). Therefore, one must conclude that equation (59) has a solution in $\overline{U}$, which ends the proof of Theorem 7.4.

Remark   It is obvious from the proof that the causality of L is essential in obtaining the existence of the resolvent operator $R = (I - L)^{-1}L$. One could drop the hypothesis on causality if we assume instead the existence, continuity and compactness of the operator $(I - L)^{-1}$. Of course, this situation may occur even in case L is a linear operator of Fredholm type in the classical sense:

$$(Lx)(t) = \int_0^T m(t, s)x(s) \, ds.$$

Corollary 7.5   *Consider the functional differential equation*

$$\dot{x}(t) = (Lx)(t) + (Nx)(t), \tag{67}$$

and initial condition $x(0) = \theta \in \mathbb{R}^n$, with L and N acting on $L^p([0, T], \mathbb{R}^n)$. It is assumed that both operators are continuous on $L^p([0, T], \mathbb{R}^n)$, with L linear and causal and N taking bounded sets into bounded sets. Moreover, it is assumed there exists a positive number M_0, such that any L^p-solution of the parametrized equation

$$\dot{x}(t) = (Lx)(t) + \lambda(Nx)(t), \quad \lambda \in (0, 1), \tag{68}$$

satisfying the initial condition $x(0) = \theta$, also satisfies $|x|_p \neq M_0$. Then, equation (67) has an absolutely continuous solution on $[0, T]$, such that $x(0) = \theta$.

Proof   Equation (67), together with the initial condition $x(0) = \theta$, is equivalent to the form (59)

$$x(t) = \int_0^t (Lx)(s) \, ds + \lambda \int_0^t (Nx)(s) \, ds. \tag{69}$$

One can rewrite (69), with obvious notation, in the form

$$x(t) = (L_1 x)(t) + \lambda(N_1 x)(t),\qquad\qquad(70)$$

with both L_1 and N_1 continuous and compact on $L^p([0, T], \mathbb{R}^n)$. Moreover, L_1 is linear and causal. The compactness of both L_1 and N_1 follows easily from the hypotheses made on L and N. Both L_1 and N_1 take $L^p([0, T], \mathbb{R}^n)$ into $AC([0, T], \mathbb{R}^n) \subset C([0, T], \mathbb{R}^n)$, and the Ascoli–Arzelà criterion applies without difficulty. Since convergence in C implies convergence in L^p, the assertion is obvious.

At this point, Theorem 7.4 applies and yields Corollary 7.5.

Let us point out that an application of the Leray–Schauder principle to the theory of classical integral equations of Hammerstein type

$$x(t) = \int_a^b k(t, s) f(s, x(s))\, ds$$

has been provided in C. Corduneanu [10], Chapter IV.

Several existence results have been obtained by the application of the Leray–Schauder fixed point principle by M. Meehan and D. O'Regan [1], including the case of equations

$$\dot{x}(t) = f(t) + \int_0^t k(t, s) g(s, x(s))\, ds,$$

for which rather handy conditions are spelled out.

Theorem 7.4 and its corollary were obtained by D. O'Regan [1]; this paper contains several other similar existence results. See also D. O'Regan [2], where equations are replaced by inclusions.

7.5   Review of further results and topics

This last section of the book is dedicated to the review of some results and topics related to the theory of functional or functional differential equations with causal operators, which present some interest in the context of this volume.

In a recent paper of L. Faina [1], the following type of *"neutral" functional differential equation* is considered:

$$\dot{x}(t) = f(t, x(t), \dot{x}(t)),\qquad\qquad(71)$$

on the whole real axis $\mathbb{R}$, with an initial condition of infinite delay type of the form

$$x(t) = \varphi(t),\quad t \in (-\infty, t_0].\qquad\qquad(72)$$

Actually, the right-hand side of (71) represents an operator, since it is assumed that

$$f : R \times C(\mathbb{R}, \mathbb{R}^n) \times L^1(\mathbb{R}, \mathbb{R}^n) \to \mathbb{R}^n.$$

Let us point out that equation (71) can be rewritten in the usual form $y(t) = (Vy)(t)$, if one denotes $y(t) = \dot{x}(t)$. Indeed, it becomes

$$y(t) = f\left(t, \varphi(t_0) + \int_{t_0}^t y(s)\, ds, y(t)\right),\qquad\qquad(73)$$

and we regard (73) as an equation in $L^1([t_0, \infty), \mathbb{R}^n)$. Of course, any solution or (73) generates a solution of (71), (72).

Assuming Carathéodory's type conditions on f, and a certain admissibility condition for the datum φ, the *local existence* of the solution for (71), (72) is obtained. Continuous dependence is also investigated and more *general hereditary structures* are discussed. For more details on these structures see also the papers by P. Brandi and R. Ceppitelli [1], R. Ceppitelli and L. Faina [1], Cristina Marcelli and Anna Salvadori [1], as well as the references therein.

The concept of *approximate solutions* to equations of the form $x(t) = (Vx)(t)$, in which x takes values in a real Banach space B and V is a causal operator on $C([0, T], B)$, is defined and discussed by L.N. Lyapin [1]. Other spaces than $C([0, T], B)$ are considered, including the L^p-spaces $L^p([0, T], B)$, $1 \leq p < \infty$. A map $x : [0, T] \to B$ is in L^p, if it is (Bochner) measurable and

$$|x|_p = \left(\int_0^T |x(s)|^p ds \right)^{1/p} < \infty. \tag{74}$$

A function $x_\varepsilon(t) \in \mathcal{C}([0, T], B)$ is called an ε-*approximate solution* $(\varepsilon > 0)$ if $|x_\varepsilon(t) - (Vx_\varepsilon)(t)|_C < \varepsilon$, for $t \in [0, T]$.

Let $U \subset \mathcal{C}([0, T], B)$ be the ball of radius $r, r > 0$, centered at the constant function $c \in \mathcal{C}([0, T], B)$. We will assume V to be causal,

$$VU \subset U, \tag{75}$$

and also that V has the fixed initial value property: $(Vx)(0) = x^0 \in B$, with x^0 independent of $x \in U$.

Then, if VU is an equicontinuous set, there exists an ε-approximate solution $x_\varepsilon(t)$ of $x(t) = (Vx)(t)$, $x_\varepsilon(t) \in U$, such that the initial condition $x_\varepsilon(0) = x^0$ is satisfied. Moreover, if V is continuous on $\mathcal{C}([0, T], B) = U$, then the set $\{x_\varepsilon(t)\}$ is connected for each fixed $\varepsilon > 0$.

A similar result is established in case the space $\mathcal{C}([0, T], B)$ is replaced by the space $L^p([0, T], B)$, $1 \leq p < \infty$.

In Chapter 4, we have seen that the solution of the linear equation $\dot{x}(t) = (Lx)(t) + f(t)$, with L a linear causal operator, admits the integral representation by means of the formula

$$x(t) = X(t, 0)x^0 + \int_0^t X(t, s) f(s) \, ds. \tag{76}$$

In particular, if we choose $x^0 = \theta$, from (76) we find the formula

$$x(t) = \int_0^t X(t, s) f(s) \, ds, \tag{77}$$

where $X(t, s)$ is the Cauchy matrix associated with the operator L. Of course, $x(t)$ given by (77) is the only solution which vanishes at $t = 0$. When we derived (76), it was assumed that the space $L^2_{\text{loc}}([0, T], \mathbb{R}^n)$ is the underlying space.

In the papers by V.G. Abdrakhamanov and Ju.N. Smolin [1], and Ju.N. Smolin [1], the problem of *integral representation* of the solution of linear equations with causal operators is discussed in the following framework.

One considers the functional linear equation

$$(Lx)(t) = f(t), \quad t \in [0, T], \tag{78}$$

where L is a linear causal operator (not necessarily continuous!) on the space $AC([0, T], \mathbb{R}^n)$, with values in $L^1([0, T], \mathbb{R}^n)$. Apparently, (78) is not a functional differential equation. But taking into account the properties of absolutely continuous functions, we realize that the following choice is possible in (78):

$$(Lx)(t) = \dot{x}(t) - (Ax)(t), \tag{79}$$

with A another causal linear operator. Therefore, given the choice of the underlying space, (78) contains the usual linear equations with causal operators.

It is assumed that the operator L on the space $AC([0, T], \mathbb{R}^n)$ is given by

$$(Lx)(t) = \frac{d}{dt} \int_0^t Q(t, s)\dot{x}(s)\, ds, \tag{80}$$

with $Q(t, s)$ satisfying certain properties and acting on L^1.

Let us point out that (80) is similar to formula (4.26). The class of operators defined by (80) is quite comprehensive and contains all continuous linear operators from AC into L^1.

The problem discussed in the above mentioned papers consists in providing conditions assuring the validity of the integral representation (77), by means of a Cauchy matrix $X(t, s)$, for the solution of equation (78), such that $x(0) = \theta$. In other words, under what conditions does the Cauchy matrix exist?

The answer is surprisingly simple. A necessary and sufficient condition for the validity of (77) is the validity of the equation

$$\int_s^t X(t, u) d_u Q(u, s) = -I, \quad 0 \le s \le t \le T, \tag{81}$$

where I stands for the unit matrix of order n.

We may also say that the integral equation (81) must have a solution $X(t, s)$, which is acceptable in (77).

The problem of the integral representation of the solution to linear equations of the form (78), with $L : AC([0, T], \mathbb{R}^n) \to L^1([0, T], \mathbb{R}^n)$ a causal operator is also treated by N.V. Azbelev, V.P. Maksimov and L.F. Rakhmatullina [1], [2]. More references can be found in these books.

In the papers of J.T. Kiguradze and Z.P. Sokhadze [1–3], *singular functional differential equations* of the form $\dot{x}(t) = (Vx)(t), t \in [0, T]$, are considered. It is assumed that

$$V : \mathcal{C}([0, T], \mathbb{R}^n) \to L^1_{\mathrm{loc}}((0, T], \mathbb{R}^n)$$

is a causal continuous operator. Instead of the usual initial condition $x(0) = x^0 \in \mathbb{R}^n$, a condition of the form

$$\lim \frac{|x(t) - x^0|}{h(t)} = 0 \quad \text{as } t \to 0+ \tag{82}$$

is imposed, where $h : [0, T] \to [0, \infty)$ is a continuous function satisfying $h(t) > 0$ for $0 < t \le T$.

In order to obtain a local existence result, the authors assume the following conditions on V and h:

For all $x(t) \in \mathcal{C}([0, T], \mathbb{R}^n)$, with $|x|_C \le r, r > 0$ being given, one has the estimate

$$\langle (V(x^0 + hx))(t), \operatorname{sgn}(x(t)) \rangle \le p(t) \sup(|x(s)|; \ 0 \le s \le t) + q(t), \tag{83}$$

a.e. on $[0, T]$, where $p(t), q(t) \in L^1([0, T], \mathbb{R}^n)$ are nonnegative, and satisfy

$$\limsup_{t \to 0+} \left(\frac{1}{h(t)} \int_0^t p(s)\, ds \right) < 1, \tag{84}$$

and

$$\lim_{t \to 0+} \left(\frac{1}{h(t)} \int_0^t q(s)\, ds \right) = 0,$$

with $\operatorname{sgn}(x) = (\operatorname{sgn} x_1, \operatorname{sgn} x_2, \dots, \operatorname{sgn} x_n)$.

Then, the (local) existence of a solution is guaranteed for the equation $\dot{x}(t) = (Vx)(t)$, under condition (82). Further results are given, concerning uniqueness, continuous dependence, and continuability (global existence) of the solution.

Among various generalizations of the concept of causal operator, we shall deal here with one that has been recently investigated by V.J. Sumin [1]. This time t will not mean a real variable but a vector in $\mathbb{R}^m : t = (t_1, t_2, \dots, t_m)$.

Let $M \subset \mathbb{R}^m$ be a bounded measurable set, and Σ a σ-algebra of measurable subsets of M. By $T \subset \Sigma$ one denotes a fixed part of Σ. $L^p(M)$ will designate the Lebesgue space of n-vector measurable maps $z(t) = \operatorname{col}(z_1(t), \dots, z_n(t)), t \in M$. By $S_k(M)$ one denotes the set of all measurable (a.e. finite) k-vector valued maps on M, with values in $\mathbb{R}^k$.

An operator $V : L^p(M) \to S_k(M)$ is called a *Volterra operator* with respect to the system of sets T, if the following property holds: for any $H \in T$ and $x, y \in L^p(M)$, from $x(t) = y(t)$ a.e. on H, there follows $(Vx)(t) = (Vy)(t)$ a.e. on H.

It is obvious that in the case of causal operators we have considered in this book, one has to take $M = [0, T'] \subset \mathbb{R}$, while T consists of all intervals $[0, t], t \le T'$.

For the generalized Volterra operators defined above, it is possible to build up an existence theory related to the equation $x(t) = (Vx)(t), t \in \mathbb{R}^m$. We refer the reader to the above mentioned reference by V.Ju. Sumin, which contains results on existence, uniqueness, continuation and approximation of solutions. More references, including some dealing with applications to control processes, can be found in V.Ju. Sumin [1].

L.M. Berezanskii [3] has investigated the exponents of solutions to linear equations (actually, of neutral type)

$$(Lx)(t) = (P\dot{x})(t) + (Tx)(t) = f(t), \quad t \in \mathbb{R}_+, \tag{85}$$

where $P, T : L^1(\mathbb{R}_+, \mathbb{R}^n) \to L^1(\mathbb{R}_+, \mathbb{R}^n)$ are linear causal operators.

The exponent is defined in terms of the Cauchy matrix $X(t, s)$ associated with (85). More precisely, it is assumed that $X(t, s)$ exists, and the solutions of (85) can be represented by the formula

$$x(t) = X(t, 0)x^0 + \int_0^t X(t, s) f(s)\, ds. \tag{86}$$

As we have seen in Chapter 4, this is always the case when $P = I =$ the identity operator, and T is continuous.

If the Cauchy matrix satisfies an estimate of the form

$$|X(t, s)| \leq N \exp\{\lambda(t - s), \ t \geq s \geq 0\}, \tag{87}$$

with $N = N(\lambda) > 0$, $\lambda \in \mathbb{R}$, then the *exponent* is $\inf\{\lambda : \lambda \in \mathbb{R}\}$, such that an estimate like (87) does hold.

The following result is established in the above quoted paper by L.M. Berezanskii: if P, T are linear bounded operators on $L^1(\mathbb{R}_+, \mathbb{R}^n)$, and P^{-1} exists, then the exponent of the system (85) is finite.

The significance of the exponent in regard to the asymptotic growth of the solutions stems from the following simple remark: if $X(t, s)$ satisfies a condition of the form (87), and f is bounded on $\mathbb{R}_+$, then (85) yields

$$|x(t)| \leq N|x^0| \exp\{\lambda t\} + NF\lambda^{-1},$$

which shows that the order of growth of the solutions of (85) is the same as the order of growth of $\exp\{\lambda t\}$.

We will conclude this section with some remarks on the topic of *partial stability* for functional differential equations with causal operators. Sometimes, not all the variables involved in the dynamics of a system have the same "weight" in describing the motion. The stability of these variables considered to be more important for the dynamics of the system leads to the concept of *partial stability*.

Let us restrict ourselves to the case of linear equations, which can be written in the form

$$\dot{x}(t) = (Ax)(t) + (By)(t),$$
$$\dot{y}(t) = (Cx)(t) + (Dy)(t). \tag{88}$$

In the system (88), $x : \mathbb{R}_+ \to \mathbb{R}^n$, $y : \mathbb{R}_+ \to \mathbb{R}^m$, while A, B, C, D are linear causal operators on certain function spaces. For instance, $A : L^\infty(\mathbb{R}_+, \mathbb{R}^n) \to L^\infty(\mathbb{R}_+, \mathbb{R}^n)$, while $B : L^\infty(\mathbb{R}_+, \mathbb{R}^n) \to L^\infty(\mathbb{R}_+, \mathbb{R}^n)$. Similarly for C and D.

The initial conditions for (88) are

$$x(t) = \varphi(t), \quad 0 \leq t < t_0, \qquad x(t_0) = x^0,$$
$$y(t) = \psi(t), \quad 0 \leq t < t_0, \qquad y(t_0) = y^0. \tag{89}$$

The zero solution $x = \theta_n$, $y = \theta_m$ of (88) is called *x-partially stable* if for any $\varepsilon > 0$ and $t_0 > 0$, there exists $\delta(\varepsilon, r_0) > 0$ such that $|x(t)| < \varepsilon$ for $t \geq t_0$, provided $|\varphi|, |\psi|, |x^0|, |y^0| < \delta(\varepsilon, t_0)$. By $|\varphi|, |\psi|$ we denote the norms in the spaces chosen as initial value spaces.

In the linear case of systems of the form (88), one can use the integral representation of solutions (see Chapter 4), express y in terms of x from the second equation and then substitute into the first. The result is a linear functional differential equation in x, of the form investigated in Chapter 5.

A theory of partial stability for functional differential equations with causal operators is not yet built up. The book by V.I. Vorotnikov [1] offers the background for developing such a theory.

Appendix

Almost periodic solutions to neutral functional equations

One problem often encountered in the applications of functional equations is the existence of periodic or almost periodic (in time) solutions. In this appendix, the hypothesis of causality we have adopted throughout this book is no longer necessary.

For basic definitions and properties concerning almost periodic functions we refer the reader to our book *Almost Periodic Functions* (Second English edition, Chelsea, New York, 1989; currently distributed by the American Mathematical Society).

We shall consider the functional equation

$$(Vx)(t) = (Wx)(t), \quad t \in \mathbb{R}, \tag{1}$$

where V and W stand for operators (not causal, in general) on the space $AP(\mathbb{R}, \mathbb{R}^n)$, consisting of Bohr almost periodic functions on $\mathbb{R}$, with values in $\mathbb{R}^n$.

Let us first assume that V in (1) is a linear operator on $AP(\mathbb{R}, \mathbb{R}^n)$, and rewrite the equation (1) in the form

$$(Lx)(t) = (Nx)(t), \quad t \in \mathbb{R}, \tag{2}$$

with N standing, in general, for a nonlinear operator on the space $AP(\mathbb{R}, \mathbb{R}^n)$. It is possible to deal with spaces of almost periodic functions other than $AP(\mathbb{R}, \mathbb{R}^n)$.

The special case of (2)

$$(Vx)(t) = f(t), \quad t \in \mathbb{R}, \tag{3}$$

with $f \in AP(\mathbb{R}, \mathbb{R}^n)$, is solvable if and only if

$$\sup |(Lx)(t)| \geq m \sup |x(t)|, \quad t \in \mathbb{R}, \tag{4}$$

for some positive m, and any $x \in AP(\mathbb{R}, \mathbb{R}^n)$. Condition (1) is the well-known condition for the invertibility (with bounded inverse) of the linear continuous operator L, taking into account that the supremum is the norm in $AP(\mathbb{R}, \mathbb{R}^n)$.

Based on the solvability of equation (3), under condition (4), we can proceed to the discussion of equation (2). It turns out that (2) is also uniquely solvable in $AP(\mathbb{R}, \mathbb{R}^n)$, if N is Lipschitz continuous, with a sufficiently small Lipschitz constant K:

$$|Nx - Ny|_{AP} \leq K|x - y|_{AP}. \tag{5}$$

Indeed, from (2) we obtain

$$m \sup |x(t) - y(t)| \leq \sup |(Lx)(t) - (Ly)(t)|$$
$$\leq \sup |(Nx)(t) - (Ny)(t)|$$
$$\leq K \sup |x(t) - y(t)|.$$

The above estimates show that the iteration process defined by

$$(Lx^m)(t) = (Nx^{m-1})(t), \quad m \geq 1, \tag{6}$$

with $x^0(t)$ arbitrary in $AP(\mathbb{R}, \mathbb{R}^n)$ is convergent in this space when $K < m$. We can obviously write

$$\sup |x^{m+1}(t) - x^m(t)| \leq Km^{-1} \sup |x^m(t) - x^{m-1}(t)|, \tag{7}$$

and since $Km^{-1} < 1$, the assertion is proved. The uniqueness of the solution of (2) is obtained from

$$m \sup |x(t) - y(t)| \leq K \sup |x(t) - y(t)|,$$

which has been established above.

The discussion conducted so far leads to the following result.

Proposition 1   *Consider equation (2) in the space* $AP(\mathbb{R}, \mathbb{R}^n)$, *and assume that the operators L and N satisfy (4), resp. (5). If $K < m$, then the iteration process defined by (6) is convergent in* $AP(\mathbb{R}, \mathbb{R}^n)$ *to the unique solution of (2).*

Remark   The existence of the solution to equations (1) and (2), can be interpreted as the existence of coincidence points to the pair of operators (V, W), resp. (L, N).

Example   As an illustration of Proposition 1, we shall consider the equation

$$x(t) + \int_{\mathbb{R}} k(t - s)x(s)ds = f(t, x(t), x(t + h)), \tag{8}$$

in which $k = (k_{ij})_{n \times m}$ is integrable on $\mathbb{R}$, while f is almost periodic in the first argument and Lipschitz continuous with respect to the last two arguments. One chooses $h \in \mathbb{R}$ arbitrarily.

The condition equivalent to (4) can be written as

$$\left| \det[I + \tilde{k}(i\omega)] \right| \geq 0, \quad \omega \in \mathbb{R}, \tag{9}$$

where

$$\tilde{k}(i\omega) = \int_{\mathbb{R}} k(t)e^{-i\omega t} dt, \quad \omega \in \mathbb{R}, \tag{10}$$

is the Fourier transform of k.

According to Proposition 1, equation (8) has a unique solution in $AP(\mathbb{R}, \mathbb{R}^n)$ if (9) is satisfied and the Lipschitz constant for f is sufficiently small.

We shall consider again equation (2), and notice that under condition (4) it can be rewritten in the form

$$x(t) = L^{-1}((Nx)(t)), \quad t \in \mathbb{R}. \tag{11}$$

In this form, application of the fixed point method appears to be appropriate. We shall make such assumptions that will allow us to obtain the existence of a solution by means of the Schauder fixed point theorem for compact operators.

In order to secure the compactness of the operator $L^{-1}N$ it suffices to assume that N is a compact operator on $\mathrm{AP}(\mathbb{R}, \mathbb{R}^n)$. Since (9) implies that $|L^{-1}| \leq m$, one can write

$$|L^{-1}(Nx)(t)| \leq m^{-1}|(Nx)(t)|, \quad t \in \mathbb{R}. \tag{12}$$

Denote

$$a(r) = \sup |(Nx)(t)|, \quad |x(t)| \leq r, \tag{13}$$

assuming, of course, that the supremum in (13) is finite for each $r > 0$. From (12) and (13) we derive

$$|L^{-1}(Nx)(t)| \leq m^{-1}a(r) \quad \text{for } |x(t)| \leq r. \tag{14}$$

Therefore, the operator $L^{-1}N$ will take the ball of radius r, centered at the zero element of $\mathrm{AP}(\mathbb{R}, \mathbb{R}^n)$, into itself, if for this r one has

$$m^{-1}a(r) \leq r. \tag{15}$$

Let us point out the fact that we need *only one value* of $r > 0$, such that (15) is valid. Such values for r do exist, for instance, if we assume

$$\limsup_{r \to \infty} \frac{a(r)}{r} < m. \tag{16}$$

In particular, when $a(r)$ grows slower than r at infinity, condition (16) is satisfied.

Summarizing the discussion carried out above, the following existence (only!) result can be stated.

Proposition 2 *Consider equation (2), with L and N continuous operators on $\mathrm{AP}(\mathbb{R}, \mathbb{R}^n)$. Moreover, assume that L is linear and invertible, while N is compact from $\mathrm{AP}(\mathbb{R}, \mathbb{R}^n)$ into itself, and such that (15) or (16) is valid. Then equation (2) has a solution in $\mathrm{AP}(\mathbb{R}, \mathbb{R}^n)$.*

Remark The compactness of a set $S \subset \mathrm{AP}(\mathbb{R}, \mathbb{R}^n)$ is equivalent to the following conditions: (a) S is bounded, i.e., there exists $M > 0$ such that $|x(t)| \leq M, t \in \mathbb{R}$, for each $x \in S$; (b) S is equi-continuous, i.e., for each $\varepsilon > 0$, there exists $\delta(\varepsilon) > 0$, such that $t, s \in \mathbb{R}, |t - s| < \delta$ implies $|x(t) - x(s)| < \varepsilon$ for any $x \in S$; (c) S is equi-almost periodic, i.e., for each $\varepsilon > 0$, there exists $\ell(\varepsilon) > 0$, such that $|x(t + \tau) - x(t)| < \varepsilon, t \in \mathbb{R}$, for at least one τ in any interval $\ell \subset \mathbb{R}$, and any $x \in S$.

Let us now return to equation (1), and consider the case similar to (3), namely

$$(Vx)(t) = f(t), \quad t \in \mathbb{R}, \tag{17}$$

in which V acts on $AP(\mathbb{R}, \mathbb{R}^n)$ and $f \in AP(\mathbb{R}, \mathbb{R}^n)$. Since V is, in general, a nonlinear operator, we do not have a condition of the form (4) to guarantee the existence of the inverse operator V^{-1}.

Following E. Zeidler (*Nonlinear Functional Analysis and Its Applications*, II, Springer, Berlin, 1983) we shall impose on the operator V in (17) a monotonicity condition:

$$m|x(t) - y(t)|^2 \leq \langle (Vx)(t) - (Vy)(t), x(t) - y(t) \rangle. \tag{18}$$

In (18), $m > 0$ is fixed, while $x, y \in AP$ are arbitrary.

As we shall see, condition (18) assures the existence of the inverse operator, which means that equation (17) is solvable in $AP(\mathbb{R}, \mathbb{R}^n)$. Actually, an iteration process can be applied in order to obtain the existence of the solution for (17).

Let us consider the auxiliary operator

$$(T_\lambda x(t) = x(t) - \lambda[(Vx)(t) - f(t)], \quad t \in \mathbb{R}, \tag{19}$$

where λ is a positive number. It is obvious that any fixed point of T_λ in $AP(\mathbb{R}, \mathbb{R}^n)$ is a solution to equation (17). It will now be shown that we can find $\lambda > 0$, such that T_λ is a contraction on $AP(\mathbb{R}, \mathbb{R}^n)$. One more condition will be required for V, namely

$$|Vx - Vy|_{AP} \leq M|x - y|_{AP}, \tag{20}$$

where $M > 0$ is fixed and $x, y \in AP(\mathbb{R}, \mathbb{R}^n)$ are arbitrary.

The following equality follows easily from the properties of the scalar product (in $\mathbb{R}^n$!):

$$|T_\lambda x - T_\lambda y|_{AP}^2 = |x - y|_{AP}^2 - 2\lambda \langle Vx - Vy, x - y \rangle + \lambda^2 |Vx - Vy|_{AP}^2,$$

which leads to the inequality

$$|T_\lambda x - T_\lambda y|_{AP}^2 \leq (1 - 2m\lambda + M^2\lambda^2)|x - y|_{AP}^2, \tag{21}$$

if we take into account (18) and (20).

From (21) we find that T_λ is a contraction if we can achieve $1 - 2m\lambda + M^2\lambda^2 < 1$ for some positive λ. This is obvious if we choose $0 < \lambda < 2mM^{-2}$. Therefore, with such λ, the operator T_λ is a contraction.

Summarizing the above discussion of equation (17), we can state the following result.

Proposition 3   *Consider equation (17), with V acting on $AP(\mathbb{R}, \mathbb{R}^n)$ and $f \in AP(\mathbb{R}, \mathbb{R}^n)$ arbitrary. If V satisfies the monotonicity condition (18), then (17) has a unique solution in $AP(\mathbb{R}, \mathbb{R}^n)$. This solution can be obtained by the iteration process $x^{m+1}(t) = (T_\lambda x^m)(t), m \geq 0, 0 < \lambda < 2mM^{-2}$, starting with an arbitrary $x^0(t) \in AP(\mathbb{R}, \mathbb{R}^n)$.*

We shall now consider the general equation (1), under the basic assumption that a solution of this equation does exist. Since this solution is automatically almost periodic, it is interesting to establish some connection between the almost-periods of the solution and those of the data.

Assume that the operator V satisfies the monotonicity condition (18). If $x(t)$ is a solution of (1), then the following inequality holds:

$$m|x(t + \tau) - x(t)|^2 \leq \langle (Wx)(t + \tau) - (Wx)(t), x(t + \tau) - x(t) \rangle. \tag{22}$$

Equation (22) leads to

$$m|x(t + \tau) - x(t)|^2 \leq |(Wx)(t + \tau) - (Wx)(t)| \cdot |x(t + \tau) - x(t)|,$$

from which we derive,

$$|x(t + \tau) - x(t)| \leq m^{-1}|(Wx)(t + \tau) - (Wx)(t)|, \tag{23}$$

for $t \in \mathbb{R}$ and τ a fixed real number.

The inequality (23) can be easily dealt with to find the connection between the almost-periods of $x(t)$ and those of the equi-almost periodic set $\{Wy\}$, where $y \in AP(\mathbb{R}, \mathbb{R}^n)$ is such that $|y(t)| \leq \sup|x(t)|$, while W is assumed compact on $AP(\mathbb{R}, \mathbb{R}^n)$. Namely, one reads from (23) that any $m\varepsilon$-almost-period for the functions in $\{Wy\}$, $|y(t)| \leq \sup|x(t)|, t \in \mathbb{R}$, is an ε-almost period for $x(t)$.

Let us point out the fact that the almost-periods of the functions in $\{Wy\}$, $|y(t)| \leq \sup|x(t)|, t \in \mathbb{R}$, depend only on the properties of the operator W.

These remarks are useful if we look for solutions of equation (1), in the form

$$\sum_k a_k \exp\{i\lambda_k t\}.$$

Finally, in concluding this appendix, let us consider an alternate approach in regard to the almost-periodicity of solutions of functional equations, such that the case of functional differential equations can be covered.

Let us assume that in equation (2) the operator L is a differential operator of the form

$$(\mathcal{L}x)(t) = \dot{x}(t) - (Lx)(t), \quad t \in \mathbb{R}. \tag{24}$$

Of course, it is necessary to choose another underlying space than $AP(\mathbb{R}, \mathbb{R}^n)$. It appears natural to consider the space $AP^{(1)}(\mathbb{R}, \mathbb{R}^n)$, consisting of all functions such that $x(t), \dot{x}(t) \in AP(\mathbb{R}, \mathbb{R}^n)$, the natural norm being $\sup(|x(t)| + |\dot{x}(t)|), t \in \mathbb{R}$. Endowed with this norm, $AP^{(1)}(\mathbb{R}, \mathbb{R}^n)$ becomes a Banach space.

The invertibility of $\mathcal{L}$, given by (24), in the space $AP^{(1)}(\mathbb{R}, \mathbb{R}^n)$ can be discussed by means of the equation $(\mathcal{L}x)(t) = f(t)$. The general case, to the best of our knowledge, has not been investigated in the literature. The case $(Lx)(t) = A(t)x(t)$ is throroughly investigated in by M.A. Krasnoselskii *et al.* in *Nonlinear Almost Periodic Oscillations*, Wiley, New York, 1973.

The case when L is a causal operator may be dealt with on the same lines as the case mentioned above. This idea can be motivated by the fact that the equation $(\mathcal{L}x)(t) = f(t)$, with L causal, possesses an integral representation of the solutions (see Chapter 3).

References

[1] V.G. Abdrakhamanov and Ju.N. Smolin, A criterion for the existence of a Green matrix in the Cauchy problem for a functional differential equation. *Russian Acad. Sci., Dokl. Math.*, **49** (1994), 174–175.

[1] E.J. Anderson, P. Nash and A.F. Perold, *Linear Programming in Infinite Dimensional Spaces*, Wiley-Interscience, Chichester, 1987.

[1] N.V. Azbelev, Some tendencies towards generalization of differential equations, *Differential Equations* (Transl.), **21** (1984), 1291–304.

[2] N.V. Azbelev, The current status and trends in development of the theory of functional differential equations, *Russian Math.* (Izv. VUZ.), **38** (1994), No. 6, 6–17.

[3] N.V. Azbelev, Recent trends in the theory of nonlinear functional differential equations, World Congress of Nonlinear Analysts '92 (V. Lakshmikantham, ed.), Walter de Gruyter (1996), 1807–1814.

[4] N.V. Azbelev, Stability and asymptotic behavior of solutions of equations with aftereffect. In the volume *Volterra Equations and Applications*, Gordon and Breach, Reading, UK, (2000).

[1] N.V. Azbelev and L.M. Berezanskii, Stability of solutions of equations with delay (Russian), *Functional Differential Equations*, Perm (1989), 3–15.

[1] N.V. Azbelev, L.M. Berezanskii, P.M. Simonov and A.V. Chistyakov, Stability of linear systems with time-lag, I, II, III, IV, *Differential Equations* (Transl.), **23** (1987), 493–500; **27** (1991), 1165–1172; **29** (1993), 153–160.

[1] N.V. Azbelev, V.P. Maksimov and L.F. Rakhmatullina, *Introduction to the Theory of Functional Differential Equations* (Russian), Moscow, Nauka (English Translation, World Fed. Publ., Atlanta, 1996) (1991).

[2] N.V. Azbelev, V.P. Maksimov, L.F. Rakhmatullina, *Contemporary Theory of Functional Differential Equations* (to appear).

[1] N.V. Azbelev and L.F. Rakhamatullina, On the problem of functional differential inequalities and monotone operators, *Functional Differential Equations* (Russian), Perm (1986), 3–9.

[2] N.V. Azbelev and L.F. Rakhamatullina, Theory of linear abstract functional differential equations and applications, *Memoirs on Differential Equations and Mathematical Physics, Georgian Academy*, **8** (1996), 1–102.

[1] N.V. Azbelev and P.M. Simonov, *Stability of Differential Equations with Aftereffect*, to appear.

[1] L.M. Berezanskii, Linear functional differential equations on a half–axis; stability of solutions (Russian), *Mat. Fizika i Nel. Mekhanika*, **4** (38), (1985), 28–34.

[2] L.M. Berezanskii, Stability criterion for differential equations with delay. *Functional Differential Equations*, Perm (1986), 21–23.

[3] L.M. Berezanskii, Exponents of solutions of linear functional–differential equations, *Differential Equations* (Transl.), **26** (1990), 657–664.

[4] L.M. Berezanskii, The positiveness of Cauchy functions and the stability of linear differential equations with after effect, *Differential Equations* (Transl.), **26** (1990), 1092–1100.

[1] L.M. Berezanskii and A.S. Larionov, Positiveness of the Cauchy matrix of a linear functional differential equation, *Differential Equations* (Transl.), **24** (1988), 1221–1230.

[1] P. Brandi and R. Ceppitelli, Existence, uniqueness and continuous dependence for hereditary differential equations, *J. Diff. Equ.* (1989), 317–339.

[1] A.L. Buhgeim, *Volterra Equations and Inverse Problems* (Russian), Nauka, Novosibirsk (1983).

[2] A.L. Buhgeim, *Introduction to the Theory of Inverse Problems* (Russian), Nauka, Novosibirsk (1988).

[1] A.I. Bulgakov, Functional differential inclusions with nonconvex Volterra operators in Tychonoff's sense, *Functional Differential Equations*, Perm (1990), 93–104.

[2] A.I. Bulgakov, Integral inclusions with nonconvex images and their applications to boundary value problems for differential inclusions, *Russian Acad. Sci., Sbornik Math.*, **77** (1994), 193–212.

[1] A.I. Bulgakov and V.P. Maksimov, Functional and fractional differential inclusions with nonconvex Volterra operators, *Differential Equations* (Transl.), **17** (1981), 881–890.

[1] T.A. Burton, *Volterra Integral and Differential Equations*, Academic Press, New York (1983).

[2] T.A. Burton, *Stability and Periodic Solutions for Ordinary and Functional Differential Equations*, Academic Press, New York (1985).

[3] T.A. Burton, Integral equations, implicit functions and fixed points, *Proc. AMS*, **124** (1996), 2383–2390.

[1] Z.B. Caljuk, Volterra functional inequalities, *Izv. Vuzov, Mat.* (1969), No. 3, 86–95.

[1] R. Ceppitelli and L. Faina, Type Volterra property in functional differential equations; Study of the Cauchy problem and extremal solutions, *Atti. Sem. Mat. Fis. Univ. Modena*, **XLI** (1993), 491–498.

[1] A.V. Chistyakov and P.M. Simonov, On the invertibility of Volterra operators in a class of invariant subspaces, *Functional Differential Equations*, Perm (1987), 63–68.

[1] S. Cinquini, Sulle equazioni funzionali del tipo di Volterra, *Rend. R. Accad. Naz. Lincei*, **17** (1933), 616–621.

[1] E.A. Coddington and N. Levinson, *Theory of Ordinary Differential Equations*, Mc Graw–Hill, New York (1955).

[1] C. Corduneanu, Problèmes globaux dans la théorie des équations intégrales de Volterra, *Ann. Mat. Pura Appl.*, **LVII** (1965), 349–363.

[2] C. Corduneanu, *Sur certaines équations fonctionnelles de Volterra*, Func. Ekv., **9** (1966), 119–127.

[3] C. Corduneanu, *Integral Equations and Stability of Feedback Systems*, Academic Press, New York (1973).

[4] C. Corduneanu, An existence theorem for functional equations of Volterra type, *Libertas Mathematica*, **6** (1986), 117–124.

[5] C. Corduneanu, Some global problems for Volterra functional differential equations. In: *Volterra Integrodifferential Equations in Banach Spaces*, (G. Da Prato and M. Iannelli, eds.), Longman, London (1989), 90–100.

[6] C. Corduneanu, Integral representation of solutions of linear abstract Volterra functional differential equations, *Libertas Mathematica*, **IX** (1989), 133–146.

[7] C. Corduneanu, Some control problems for abstract Volterra functional differential equations, *Proc. Int. Symp. MTNS–89*, **III** (1990), 331–338, Birkhäuser Basel.

[8] C. Corduneanu, Perturbation of linear abstract Volterra equations, *J. Integral Eqs. and Appl.*, **2** (1990), 393–401.

[9] C. Corduneanu, Abstract Volterra equations and weak topologies. In: *Lecture Notes Math. 1475*, Springer-Verlag (1991), 110–115.

[10] C. Corduneanu, *Integral Equations and Applications*, Cambridge University Press (1991).

[11] C. Corduneanu, Equations involving abstract Volterra operators, *Functional Differential Equations* (J. Kato and T. Yoshizawa, eds.), World Scientific, Singapore (1991), 55–66.

[12] C. Corduneanu, *Kneser property for abstract functional differential equations of Volterra type*, World Scientific Series in Applied Mathematical Analysis, **I** (1992), 111–118.

[13] C. Corduneanu, An abstract LQ-optimal control problem and its applications, *Libertas Mathematica*, **XII** (1992), 21–27.

[14] C. Corduneanu, LQ-optimal control problems for systems with abstract Volterra operators, *Tekhn. Kibernetika* (1993), No. 1, 132–136.

[15] C. Corduneanu, Equations with abstract Volterra operators as modelling tools in Science and Engineering, *Integral Methods in Science and Engineering* (C. Constanda, ed.), Longman London (1994).

[16] C. Corduneanu, Functional differential equations with abstract Volterra operators and their control, In: *Ordinary Differential Equations and Their Applications*, Bologna (1995), 61–81.

[17] C. Corduneanu, Stability problems for Volterra functional differential equations, In: *Comparison Methods and Stability Theory* (X. Liu and D. Siegel, eds.), M. Dekker, New York (1995), 87–99.

[18] C. Corduneanu, Some new trends in Liapunov's second method, World Congress of Nonlinear Analysis (V. Lakshmikantham, ed.), Walter de Gruyter, Berlin, New York (1996).

[19] C. Corduneanu, Neutral functional differential equations with abstract Volterra operators. In: *Advances in Nonlinear Dynamics*, Gordon & Breach (1997), 229–235.

[20] C. Corduneanu, Neutral functional equations of Volterra type, *Functional Differential Equations* (Israel) (1997), No. 3–4, 265–270.

[21] C. Corduneanu, Abstract Volterra equations: A survey, *Math. and Comp. Modeling*, 32 (2000), 1503–1528.

[22] C. Corduneanu, Stability problems for systems of nuclear reactors. Ordinary and Delay Differential Equations. Pitman Research, Notes in Mathematics, 272 (1992), 29–33.

[23] C. Corduneanu, *Principles of Differential and Integral Equations*, Chelsea, New York (1988).

[1] C. Corduneanu and V. Lakshmikantham, Equations with infinite delay: A survey, *J. Nonlinear Analysis*, TMA, **4** (1981), 831–877.

[1] C. Corduneanu and Yizeng Li, On the exponential asymptotic stability for functional differential equations with causal operators (to appear).

[1] C. Corduneanu and M. Mahdavi, Asymptotic behaviour of systems with abstract Volterra operators. In: *Qualitative Problems for Differential Equations and Control Theory* (C. Corduneanu, ed.), World Scientific, Singapore (1995).

[2] C. Corduneanu and M. Mahdavi, On neutral functional differential equations with causal operators. *Proc. of the Third Workshop of the Intern. Inst. General Systems Science*, Tianjin, China (1998), 43–48.

[3] C. Corduneanu and M. Mahdavi, On neutral functional differential equations with causal operators, II. In: *Integral Methods in Science and Engineering*, Chapman & Hall CEC Press, Res. Notes 418 (2000), 102–106.

[1] C. Corduneanu and S.Q. Zhu, *Continuity of parametrized linear–quadratic optimal control problem*. In: *Optimal Control of Differential Equations* (N.H. Pavel, ed.), M. Dekker, New York (1994).

[1] J.M. Cushing, Admissible operators and solutions of perturbed operator equations, *Funk. Ekv.*, **19** (1976), 79–84.

[2] J.M. Cushing, Strongly admissible operators and Banach space solutions of nonlinear equations, *Funk. Ekv.*, **20** (1977), 237–245.

[1] K. Deimling, *Functional Analysis*, Springer-Verlag, Berlin (1985).

[1] A.I. Domoshnitzky and M.V. Sheina, Non-negativeness of Cauchy matrix and stability of systems of linear differential equations with retarded argument, *Differential Equations* (Transl.), **25** (1989), 145–150.

[1] R.D. Driver, Existence and stability of solutions of a delay differential system, *Archive Rational Mechanics Analysis*, **10** (1962), 401–426.

[1] N. Dunford and J.T. Schwarz, *Linear Operators*, Part I, Interscience, New York (1957).

[1] R.E. Edwards, *Functional Analysis: Theory and Applications*, Holt, Rinehart and Winston, New York (1965).

[1] L. Faina, Existence and continuous dependence for a class of neutral functional differential equations, *Annales Polonici Math.*, **LXIV** (1996), 215–226.

[1] L.G. Fedorenko, Stability of functional differential equations, *Differential Equations* (Translation from Russian), **21** (1986), 1031–1037.

[1] A. Friedman, *Foundations of Modern Analysis*, Dover, New York (1982).

[1] J. Gohberg and S. Goldberg, *Basic Operator Theory*, Birkhäuser, Boston (1981).

[1] D. Graffi, Sopra una equazione funzionale e la sua applicazione a un problema di fisica ereditaria, *Annali Mat. Pura Appl.*, **9** (1931), 143–179.

[1] G. Gripenberg, S.O. Londen and O. Staffans, *Nonlinear Volterra Integral and Functional Equations*, Cambridge University Press (1990).

[1] S.A. Gusarenko, On a generalization of the notion of Volterra operator, *Soviet Math. Dokl.*, **36** (1988), 156–159.

[2] S.A. Gusarenko, On equations with Volterra operators (manuscript).

[1] A. Halanay, *Differential Equations: Stability, Oscillations, Time Lags*, Academic Press, New York, 1966.

[1] J.K. Hale, *Theory of Functional Differential Equations*, Springer Verlag, Berlin (1977).

[1] J.K. Hale and M.A. Cruz, Existence, uniqueness and continuous dependence for hereditary systems, *Ann. Mat. Pura. Appl.*, Series 4, **85** (1970), 63–81.

[1] J.K. Hale and J. Kato, Phase space for retarded equations with infinite delay, *Funkcialaj Ekvacioj*, **4** (1978), 11–41.

[1] T. Hara and R. Miyazaki, Equivalent conditions for stability of a Volterra integro–differential equation. *J. Math. An. Appl.*, **174** (1993), 298–316.

[1] Ph. Hartman, *Ordinary Differential Equations*, Wiley, New York (1964).

[1] Y. Hino, S. Murakami and T. Naito, *Functional Differential Equations with Infinite Delay*, Springer Lecture Notes, 1473, Berlin (1991).

[1] Z. Kamont and M. Kwapisz, On the Cauchy problem for differential–delay equations in a Banach space, *Math. Nachr.*, **74** (1976), 173–190.

[1] L.V. Kantorovich and G.P. Akilov, *Functional Analysis* (Second edition), Pergamon Press, Oxford (1982).

[1] G. Karakostas, Causal operators and topological dynamics, *Ann. Math. Pura Appl.*, Series 4, **131** (1982), 1–27.

[1] S.G. Karnishin, On the stability of solution of linear functional differential equations with respect to a part of the variables. *Functional Differential Equations*, Perm (1987), 48–52.

[1] J. Kato, On a Liapunov–Razumikhin type theorem for functional differential equations. *Funk. Ekvacioj*, **16** (1973), 225–239.

[2] J. Kato, *Phase space for functional differential equations*, Colloquia Mathematica Societatis J. Bolyai, **53** (1988), 307–325.

[1] I.T. Kiguradze and Z.P. Sokhadze, On the uniqueness of a solution to the Cauchy problem for functional differential equations. *Differential Equations* (Transl.), **31** (1995), 1947–1958.

[2] I.T. Kiguradze and Z.P. Sokhadze, On singular functional differential inequalities, *Georgian Math. J.*, **4** (1997), 259–278.

[3] I.T. Kiguradze and Z.P. Sokhadze, On Cauchy's problem for singular evolutionary functional differential equations, *Differential Equations* (Transl.), **33** (1997), 48–59

[1] A.V. Kim, *Lyapunov's Direct Method in the Theory of Stability for Delay Systems* (Russian), Ekaterinburg, Ural University Press (1992).

[1] V. Kolmanovskii and A.D. Myshkis, *Applied Theory of Functional Differential Equations*, Kluwer, Dordrecht (1992).

[1] V.G. Kurbatov, Linear functional differential equations of neutral type and retarded spectrum (Russian), *Sibirsk. Math. Zh.*, **16** (1975), 438–550.

[2] V.G. Kurbatov, Stability of functional differential equations, *Differential Equations* (Transl.), **17** (1981), 963–972.

[3] V.G. Kurbatov, Stability of functional differential equations on the whole real line and on a half-line, *Differential Equations* (Translation from Russian), **22** (1986), 641–644.

[4] V.G. Kurbatov, *Linear Differential Difference Equations* (Russian), Voronezh University Press (1990).

[5] V.G. Kurbatov, Stability of neutral type equations in differential phase space, *Functional Differential Equations* (Israel), **3** (1995), 99–133.

[1] J. Kwapisz, Weighted norms and Volterra integral equations in L^p-space, *J. Appl. Math. Stochastic An.*, **4** (1991), 161–164.

[1] M. Kwapisz, Bielecki's method; existence and uniqueness results for Volterra integral equations in L^p-space, *J. Math. An. Appl.*, **154** (1991), 403–416.

[2] M. Kwapisz, Remarks on the existence and uniqueness of solutions of Volterra functional equations in L^p-space, *J. Integral Equations Appl.*, **3** (1991), 383–392.

[1] M. Kwapisz and J. Turo, Some integral–functional equations, *Funkc. Ekvacioj*, **18** (1975), 107–162.

[1] V. Lakshmikantham, Binggen Zhang and Lizhi Wen, *Theory of Differential Equations with Unbounded Delay*, Kluwer, Dordrecht (1994).

[1] V. Lakshmikantham, S. Leela and A.A. Martynyuk, *Stability of Motion by the Comparison Method* (Russian), Naukova Dumka, Kiev (1991).

[1] V. Lakshmikantham and M.R. Mohana Rao, *Theory of Integrodifferential Equations*, Gordon & Breach, Reading, UK (1995).

[1] T. Lalescu, *Introduction à la théorie des équations intégrales*, Hermann, Paris (1912).

[1] J.J. Levin, A nonlinear Volterra equation not of convolution type, *J. Diff. Equations*, **4** (1968), 176–186.

[1] Yizeng Li, Global existence and stability of functional differential equations with abstract Volterra operators. PhD Thesis, University of Texas at Arlington (1993).

[1] Yizeng Li, Abstract Volterra equations with initial point data, *Libertas Mathematica*, **XIII** (1993), 141–153.

[1] Yizeng Li, Existence and integral representation of solutions of the second kind initial value problem for functional differential equations with abstract Volterra operator, *Nonlinear Studies*, **3** (1996), 35–48.

[1] Yizeng Li, Stability problems of functional differential equations with abstract operator, *J. Integral Equations and Appl.*, **8** (1996), 47–63.

[1] Yizeng Li and Mehran Mahdavi, Linear and quasilinear equations with abstract Volterra operators. In: *Volterra Equations and Applications*, Gordon & Breach, Reading, UK (2000).

[1] X. Luo and D. Bertsimas, A new algorithm for state-constrained separated continuous linear programs. *SIAM J. Control and Optimization*, **37** (1998), 177–210.

[1] L.N. Lyapin, Volterra equations in Banach space, *Differential Equations* (Transl.), **19** (1983), 801–808.

[1] M. Mahdavi, *Contribution to the theory of functional differential equations involving abstract Volterra operators*, PhD Dissertation, University of Texas at Arlington (1992).

[2] M. Mahdavi, Nonlinear boundary value problem involving abstract Volterra operators, *Libertas Mathematica*, **XIII** (1993), 17–26.

[3] M. Mahdavi, Linear functional differential equations with abstract Volterra operators, *Diff. Integral Equations*, **8** (1995), 1517–1523.

[1] V.P. Maksimov, Linear functional differential equations (Russian), Abstract of PhD Thesis, University of Kazan (1974).

[1] A.V. Malygina, Some stability criteria for scalar linear delay equations. *Functional Differential Equations* (Russian), Perm (1992), 27–131.

[1] Ja.D. Mamedov, S. Ashirov and S. Atdaev, *Theorems on Inequalities* (Russian), *Acad. Sci. Turkm. Sov. Rep.* (1980), Ashkhabad.

[1] C. Marcelli and A. Salvadori, Interaction among the theories of ordinary differential equations in various hereditary settings, *Nonlinear Analysis – TMA*, **35** (1999), 1001–1017.

[1] G. Marinescu, Différentielles de Gâteaux et Fréchet dans les espaces localement convexes. *Bull. Math. Soc. Sci. Math. Phys. Roumanie* (N.S.), **1** (49), 1957, 77–86.

[1] A.A. Martynyuk, *Integral Inequalities in Stability Theory* (Russian), Kiev.

[1] J.L. Massera and J.J. Schäffer, *Linear Differential Equations and Function Spaces*, Academic Press, New York (1966).

[1] L. Máté, On the continuity of causal operators and the Pták–Stein theorem, *Per. Math. Hungarica*, **20** (1989), 219–230.

[1] A. McNabb and G. Weir, Comparison theorems for causal functional differential equations, *Proc. AMS*, **104** (1988), 449–452.

[1] E.J. McShane, *Integration*, Princeton University Press (1946).

[1] M. Meehan and D. O'Regan, Existence theory for nonlinear Volterra integrodifferential and integral equations, *Nonlinear Analysis*, TMA, **31** (1998), 317–341.

[1] R.K. Miller, *Nonlinear Volterra Integral Equations*, W.A. Benjamin, Menlo Park, California (1971).

[1] S. Murakami, Exponential stability for fundamental solution of some functional differential equations. In: *Functional Differential Equations* (T. Yohizawa and J. Kato, eds.), World Scientific, Singapore (1991).

[1] S. Murakami, Asymptotic behavior of solutions of some differential equations, *J. Math. An. Appl.*, **109** (1985), 534–545.

[1] A.D. Myshkis, *Linear Differential Equations with Retarded Argument* (Russian), Moscow, Nauka (1972).

[1] L. Neustadt, *Optimization (A theory of necessary conditions)*, Princeton University Press (1976).

[1] M.N. Oğuztörelli, *Time–Lag Control Systems*, Academic Press, New York (1966).

[1] Z. Okonkwo, Two examples in LQ-optimal control theory, *Libertas Mathematica*, **XIII** (1993), 177–182.

[1] A.C. Pipkin, *A Course on Integral Equations*, Springer Verlag, Berlin, 1991.

[1] B.S. Razumikhin, *Stability of Hereditary Systems* (Russian), Moscow, Nauka (1983).

[1] D. O'Regan, Abstract Volterra equations, *Pan Am. Math. J.*, **7** (1997), 19–28.

[1] D. O'Regan, Abstract operator inclusion, *Functional Differential Equations* (Israel), **4** (1997), 143–154.

[1] D. O'Regan and R. Precup, *Theorems of Leray–Schauder Type and Applications* (to appear).

[1] M. Reichert, Zuzammenhang der Fixpunktmenge bei verdichtenden Volterra–Operatoren, *Archivum Math.*, **55** (1990), 166–172.

[1] W.J. Rugh, *Nonlinear System Theory: The Volterra/Wiener Approach*, Johns Hopkins University Press, Baltimore, MD (1981).

[1] Cora Sadosky, *Interpolation of Operators and Singular Integrals*, M. Dekker, New York (1979).

[1] I.W. Sandberg, Expansions for nonlinear systems, and Volterra expansions for time-varying nonlinear systems. *Bell Syst. Techn. J.*, **61** (1982), 159–225.

[1] I.W. Sandberg, Nonlinear input–output maps and approximate representations, *AT & T Technical J.*, **64** (1985), 1967–1978.

[2] I.W. Sandberg, Uniform approximation and the circle criterion, *IEEE Trans. AC*, **38** (1993), 1450–1458.

[1] G. Sansone and R. Conti, *Nonlinear Differential Equations*, Pergamon Press, New York (1964).

[1] M. Schetzen, *The Volterra and Wiener Theories of Nonlinear Systems*, Wiley, New York (1980).

[1] P.M. Simonov and A.V. Chistyakov, Theorems on uniform exponential stability of delay equations. *Functional Differential Equations*, Perm (1991), 83–95.

[1] Ju.N. Smolin, On Cauchy's matrix for functional differential equations. (Russian), *Izv. Vyssh. Uč. Zav., Mat.* (1988), 54–62.

[1] E.M. Solnechnyi, Conditions for the Volterra property of the operator generated by the Cauchy problem for a partial differential equation. *Differential Equations* (Transl.), **33** (1997), 267–174.

[1] O.J. Staffans, A direct Lyapunov approach to Volterra integro-differential equations. *SIAM J. Math. An.*, **19** (1988), 879–901.

[1] V.I. Sumin, Functional-operator Volterra equations in the theory of optimal control with distributed parameters, *DANSSR*, **305** (1989), 1056–1059.
[2] V.I. Sumin, On functional Volterra equations, *Russian Math.* (Iz. VUZ), **39** (1995), No. 9, 67–77.
[3] V.I. Sumin, On Volterra equations in Banach spaces. *Funkc. Ekvacioj*, **20** (1977), 247–258.
[1] L. Tonelli, Sulle equazioni funzionali di Volterra, *Bull. Calcutta Math. Soc.*, **20** (1930), 31–48.
[1] F. Tricomi, *Integral Equations*, Interscience, New York (1957).
[1] M. Turinici, Abstract monotone mappings and applications to functional differential equations, *Rend. Accad. Naz. Lincei. Sc. Fis., Mat. Nat.*, **66**, (1979), 189–193.
[2] M. Turinici, Multivalued contractions and applications to functional differential equations, *Acta Math. Acad. Sci. Hungaricae*, **37** (1981), 147–151.
[1] A.N. Tychonoff, Sur les équations fonctionnelles de Volterra et leurs applications à certains problèmes de la Physique Mathématique, *Bull. Math. Univ. Moscou* (Série Intern.), **A1** (1938), No. 8.
[1] V.A. Tyshkevich, *Some Problems in Stability of Functional Differential Equations* (Russian), Kiev, Naukova Dumka (1981).
[1] V. Volterra, *Leçons sur les Équations Intégrales et les Équations Intégro-Différentielles*, Gauthier-Villars, Paris (1913).
[1] V.I. Vorotnikov, *Partial Stability and Control*, Birkhaüser, Boston, 1998.
[1] P.P. Zabreyko, A.Z. Koshelev, M.A. Krasnoselskii, S.G. Mikhlin, L.S. Rakovshchik and V. Yu. Stetsenko, *Integral Equations. A Reference Text*, Noordhoff, Leyden, 1975.
[1] L.A. Zhivotovskii, Existence theorems and uniqueness classes of solutions of functional equations with heredity, *Differential Equations* (Transl.), **7** (1971), 1043–1049.
[2] L.A. Zhivotovskii, Continuability of the solutions of functional equations with heredity. *Diff. Equations* (Transl.), **8** (1972), 2163–2166.
[1] E.S. Zhukovskii, On the theory of Volterra equations, *Differential Equations* (Transl.), **25** (1989), 1132–1137.
[2] E.S. Zhukovskii, Volterra property and spectral properties of the operator of inner superposition, *Differential Equations* (Transl.), **30** (1994), 229–234.

Index

Other titles in the Stability and Control: Theory, Methods and Applications series.